National Spatial Data Infrastructure Partnership Programs

RETHINKING THE FOCUS

Mapping Science Committee
Board on Earth Sciences and Resources
Division on Earth and Life Studies
National Research Council

NATIONAL ACADEMY PRESS
Washington, D.C.

NOTICE: The project that is the subject of this report was approved by the Governing Board of the National Research Council, whose members are drawn from the councils of the National Academy of Sciences, the National Academy of Engineering, and the Institute of Medicine. The members of the committee responsible for the report were chosen for their special competences and with regard for appropriate balance.

Support for the activities of the Mapping Science Committee is provided by the National Imagery and Mapping Agency (NMA202-99-1-1018), the U.S. Geological Survey (99HQAG0193), the Federal Geographic Data Committee, the Bureau of Transportation Statistics (DTTS59-99-P-00257), the National Oceanic and Atmospheric Administration (56-DKNA-0-95106), and the Bureau of the Census (43-YA-BC-037424). Any opinions, findings, conclusions, or recommendations expressed in this document are those of the authors and should not be interpreted as necessarily representing the views of the agencies that provided support for this project, or of the official policies, either expressed or implied, of the U.S. Government.

International Standard Book Number (ISBN) 0-309-07645-5

Additional copies of this report are available from: National Academy Press 2101 Constitution Avenue, NW Box 285 Washington, DC 20055 800–624–6242 202–334–3313 (in the Washington Metropolitan Area) http://www.nap.edu

Cover: Aerial photo courtesy of the U.S. Geological Survey. The image on the bottom of the page was dowloaded from PhotoDisc.

Copyright 2001 by the National Academy of Sciences. All rights reserved.

Printed in the United States of America

THE NATIONAL ACADEMIES

National Academy of Sciences
National Academy of Engineering
Institute of Medicine
National Research Council

The **National Academy of Sciences** is a private, nonprofit, self-perpetuating society of distinguished scholars engaged in scientific and engineering research, dedicated to the furtherance of science and technology and to their use for the general welfare. Upon the authority of the charter granted to it by the Congress in 1863, the Academy has a mandate that requires it to advise the federal government on scientific and technical matters. Dr. Bruce M. Alberts is president of the National Academy of Sciences.

The **National Academy of Engineering** was established in 1964, under the charter of the National Academy of Sciences, as a parallel organization of outstanding engineers. It is autonomous in its administration and in the selection of its members, sharing with the National Academy of Sciences the responsibility for advising the federal government. The National Academy of Engineering also sponsors engineering programs aimed at meeting national needs, encourages education and research, and recognizes the superior achievements of engineers. Dr. Wm. A. Wulf is president of the National Academy of Engineering.

The **Institute of Medicine** was established in 1970 by the National Academy of Sciences to secure the services of eminent members of appropriate professions in the examination of policy matters pertaining to the health of the public. The Institute acts under the responsibility given to the National Academy of Sciences by its congressional charter to be an adviser to the federal government and, upon its own initiative, to identify issues of medical care, research, and education. Dr. Kenneth I. Shine is president of the Institute of Medicine.

The **National Research Council** was organized by the National Academy of Sciences in 1916 to associate the broad community of science and technology with the Academy's purposes of furthering knowledge and advising the federal government. Functioning in accordance with general policies determined by the Academy, the Council has become the principal operating agency of both the National Academy of Sciences and the National Academy of Engineering in providing services to the government, the public, and the scientific and engineering communities. The Council is administered jointly by both Academies and the Institute of Medicine. Dr. Bruce M. Alberts and Dr. Wm. A. Wulf are chairman and vice chairman, respectively, of the National Research Council.

www.national-academies.org

MAPPING SCIENCE COMMITTEE

DAVID J.COWEN, *Chair,* University of South Carolina, Columbia
ANNETTE J.KRYGIEL, *Vice Chair,* Integro, Great Falls, Virginia
ERIC A.ANDERSON, Des Moines, Iowa
CLIFFORD A.BEHRENS, Telcordia Technologies, Morristown, New Jersey
WILLIAM J.CRAIG, The University of Minnesota, Minneapolis
MARK MONMONIER, Syracuse University
JOEL L.MORRISON, Ohio State University, Columbus
SHERYL G.OLIVER, Illinois Department of Natural Resources, Springfield
HARLAN J.ONSRUD, University of Maine, Orono
C.STEPHEN SMYTH, Microsoft Corporation, Redmond, Washington
JAMES V.TARANIK, University of Nevada, Reno
REX W.TRACY, BAE SYSTEMS, San Diego, California
A.KEITH TURNER, Colorado School of Mines, Golden

National Research Council Staff

THOMAS M.USSELMAN, Study Director (until 2/2000)
DAVID A.FEARY, Study Director (from 3/2000)
JENNIFER T.ESTEP, Administrative Associate
REBECCA E.SHAPACK, Research Assistant (until 3/2001)
SHANNON L.RUDDY, Project Assistant (from 3/2001)

BOARD ON EARTH SCIENCES AND RESOURCES

RAYMOND JEANLOZ, *Chair,* University of California, Berkeley
JOHN J.AMORUSO, Amoruso Petroleum Company, Houston, Texas
PAUL BARTON, JR., U.S. Geological Survey *(emeritus),* Reston, Virginia
DAVID L.DILCHER, University of Florida, Gainesville
BARBARA L.DUTROW, Louisiana State University, Baton Rouge
ADAM M.DZIEWONSKI, Harvard University, Cambridge, Massachusetts
WILLIAM L.GRAF, Arizona State University, Tempe
GEORGE M.HORNBERGER, University of Virginia, Charlottesville
SUSAN KIEFFER, Kieffer and Woo, Inc., Palgrave, Ontario
DIANNE R.NIELSON, Utah Department of Environmental Quality, Salt Lake City
JONATHAN PRICE, Nevada Bureau of Mines and Geology, Reno
BILLIE L.TURNER II, Clark University, Worcester Massachusetts

National Research Council Staff

ANTHONY R.DE SOUZA, Director
TAMARA L.DICKINSON, Senior Program Officer
DAVID A.FEARY, Senior Program Officer
ANNE M.LINN, Senior Program Officer
PAUL M.CUTLER, Program Officer
LISA M.VANDEMARK, Program Officer
KRISTEN L.KRAPF, Research Associate
KERI H.MOORE, Research Associate
MONICA R.LIPSCOMB, Research Assistant
JENNIFER T.ESTEP, Administrative Associate
VERNA J.BOWEN, Administrative Assistant
YVONNE FORSBERGH, Senior Project Assistant
KAREN IMHOF, Senior Project Assistant
SHANNON L.RUDDY, Project Assistant
TERESIA K.WILMORE, Project Assistant
WINFIELD SWANSON, Editor

Acknowledgment of Reviewers

This report has been reviewed in draft form by individuals chosen for their diverse perspectives and technical expertise, in accordance with procedures approved by the NRC's Report Review Committee. The purpose of this independent review is to provide candid and critical comments that will assist the institution in making its published report as sound as possible and to ensure that the report meets institutional standards for objectivity, evidence, and responsiveness to the study charge. The review comments and draft manuscript remain confidential to protect the integrity of the deliberative process. We wish to thank the following individuals for their reviews of this report:

Richard J.Aspinall, Department of Earth Sciences, Montana State University, Bozeman

Don F.Cooke, Geographic Data Technologies, Lebanon, New Hampshire

Stephen D.DeGloria, Department of Crop and Soil Sciences, Cornell University, Ithaca, New York

Dennis B.Goreham, National States Geographic Information Council, Salt Lake City, Utah

Matt Hoobler, Department of Agriculture, Cheyenne, Wyoming

William E.Huxhold, School of Architecture and Urban Planning, University of Wisconsin-Milwaukee

Kevin Kryzda, Information Systems Department, Martin County, Stuart, Florida

Susan Lambert, Office of Geographic Information, Frankfort, Kentucky

Although the reviewers listed above have provided many constructive comments and suggestions, they were not asked to endorse the conclusions or recommendations nor did they see the final draft of the report before its release. The review of this report was overseen by Frederick J.Doyle, U.S. Geological Survey *(emeritus)*, appointed by the Division on Earth and Life Studies, who was responsible for making certain that an independent examination of this report was carried out in accordance with institutional procedures and that all review comments were carefully considered. Responsibility for the final content of this report rests entirely with the authoring committee and the institution.

Preface

The Mapping Science Committee serves as a focus for external advice to federal agencies on scientific and technical matters related to spatial data handling and analysis. One of the committee's roles is to provide advice on the development of a robust national spatial data infrastructure for making informed decisions at all levels of government and throughout society in general.

The concept of a National Spatial Data Infrastructure (NSDI) was first advanced by the Mapping Science Committee (MSC) in its 1993 report, *Toward a Coordinated Spatial Data Infrastructure for the Nation* (NRC, 1993). The next year, the committee addressed partnerships as an essential component of the NSDI (*Promoting the National Spatial Data Infrastructure Through Partnerships;* NRC, 1994). Since then, the Federal Geographic Data Committee (FDGC) has sponsored a series of annual competitions for grants to promote the NSDI. These grants have been used to stimulate the development of partnerships at a variety of levels (local, state, federal); to encourage the documentation of data according to national standards to facilitate their sharing; and to encourage the use of geospatial data in new applications.

By 2000, these FGDC programs had provided support for NSDI development in 49 of the 50 states. Their objectives had varied substantially from year to year, and from program to program. As one of the advisory bodies responsible for originating the NSDI, the MSC identified the need for an assessment of progress to date, and for guidance on directions for the future. Was the NSDI developing according to plan, with FGDC partnership programs working to advance its goals, or was some degree of redirection appropriate? Accordingly, a study to address these questions was conducted by the MSC as one of its core activities in the latter half of 1999 and through 2000. This report is the outcome of that process. It is important to recognize that the committee focused on the partnership programs promoted by the FGDC, and has not attempted a comprehensive analysis of all NSDI partnership activities.

In addition to the present members of the MSC, I wish to acknowledge the input of former MSC members who contributed to earlier versions of the report —Brian Berry, Nick Chrisman, David Coleman, Hank Garie, Barry Glick, Karen Siderelis, and Lyna Wiggins. I would particularly like to acknowledge my predecessor as MSC chair, Mike Goodchild, who oversaw the conception of this report and made a major contribution to its content.

David J.Cowen
Chair, Mapping Science Committee

Contents

	EXECUTIVE SUMMARY	1
1	**NSDI AND PARTNERSHIPS**	5
	Goals of the NSDI	5
	Coordination and Leadership	9
	Components of the NSDI	11
	Data Standards	11
	National Geospatial Data Clearinghouse	12
	NSDI Framework	13
	Purpose of the Report	15
2	**REVIEW OF NSDI PARTNERSHIP PROGRAMS**	17
	NSDI Cooperative Agreements Program	19
	Framework Demonstration Projects Program	23
	"Don't Duck Metadata"	25
	Community Demonstration Projects	26
	Community-Federal Information Partnerships	29
	Priming the Pump—the Federal Role in NSDI Partnership Initiation	30
	The Future Federal Role in Developing the NSDI	35
3	**FUTURE PARTNERSHIPS AND THE EVOLUTION OF NSDI ACTIVITIES**	39
	Framework Data Production	40
	Data Access, Use, and Other Framework Issues	44
	The Time Dimension: Data Update, Archiving, and Change Detection	48
	Privacy, the Private Sector, and Public Access Issues	50
	The GeoData Alliance—an Innovative Organizational Approach for Development of the NSDI	53

4	**AN EXTENDED NATIONAL SPATIAL DATA INFRASTRUCTURE FRAMEWORK: THE ROLE OF OTHER ORGANIZATIONS**	57
	Arguments for an Extended Framework	57
	Definition of a City or County Extended Framework	64
	Definition of a State or Tribal Nation Extended Framework	66
	Summary of Spatial Data Themes	68
	Roles of Private Industry and Non-Profit Organizations	70
5	**CONCLUSIONS AND RECOMMENDATIONS**	73
	REFERENCES	77
	ACRONYMS	81

Executive Summary

The National Spatial Data Infrastructure (NSDI) was envisioned as a way of enhancing the accessibility, communication, and use of geospatial data to support a wide variety of decisions at all levels of society. The goals of the NSDI are to reduce redundancy in geospatial data creation and maintenance, reduce the costs of geospatial data creation and maintenance, improve access to geospatial data, and improve the accuracy of geospatial data used by the broader community. At the core of the NSDI is the concept of partnerships, or collaborations, between different agencies, corporations, institutions, and levels of government. In a previous report, the Mapping Science Committee (MSC) defined a partnership as "…a joint activity of federal and state agencies, involving one or more agencies as joint principals focusing on geographic information" (NRC, 1994; p. 19). The concept of partnerships was built on the foundation of shared responsibilities, shared costs, shared benefits, and shared control. Partnerships are designed to share the costs of creation and maintenance of geospatial data, seeking to avoid unnecessary duplication, and to make it possible for data collected by one agency at a high level of spatial detail to be used by another agency in more generalized form. Over the past seven years, a series of funding programs administered by the Federal Geographic Data Committee (FGDC) has stimulated the creation of such partnerships, and thereby promoted the objectives of the NSDI, by raising awareness of the need for a coordinated national approach to geospatial data creation, maintenance, and use. They include the NSDI Cooperative Agreements Program, the Framework Demonstration Projects Program, the Community Demonstration Projects, and the Community-Federal Information Partnerships proposal. This report assesses the success of the FGDC partnership programs (see Box) that have been established between the federal government and state and local government, industry, and academic communities in promoting the objectives of the National Spatial Data Infrastructure.

As the NSDI is explicitly a national concept, the committee considers that it is appropriate that the federal government originated and continues to play the major role in its construction. As the primary sponsors of the first stage of adoption of the NSDI, the federal government has successfully "primed the NSDI pump." This priming action appears to have been directed largely at the one specific goal of improved access to data, and the evidence gathered by the committee clearly demonstrates that the NSDI does indeed improve access to data. The actions of the federal sponsors of the NSDI, in creating the National Geospatial Data Clearinghouse (NGDC) and fostering the use of the Content Standards for Digital Geospatial Metadata (CSDGM) through partnership programs, have led to a substantial improvement in nationwide access to geospatial data. Therefore, in the data access area, we anticipate that a second stage of adoption will follow; namely, where many more agencies and organizations can be expected to participate in the NGDC and adopt the metadata standard, without requiring further direct pump-priming and encouragement by the federal government.

Full adoption of the NSDI will require attention to the remaining three goals: reduced redundancy, decreased cost, and increased accuracy. To date, the funding incentives established by the FGDC through the NSDI partnership programs do not appear to have significantly affected these goals. The committee strongly suggests that the FGDC direct its attention to the remaining three goals, in order to assure the future of the NSDI, with the understanding that successfully attaining these additional goals will require a much more fundamental level of cooperation among partners than the simple sharing of an agency's existing data. Specifically, future partnership programs sponsored by the federal government should be based on convincing evidence that adoption of the NSDI's concepts and design result in reductions in redundancy and cost, as well as increased accuracy. It will also be important that future funding initiatives be widely advertised, with the criteria for selection clearly stated. Ideally, a panel of experts in the field should evaluate the proposals, with appropriate peer-review. In an environment where programs designed to promote the NSDI may become convolved with other programs, be diverted to serve other needs, or be expected to serve too many different purposes, it is particularly important that a program of partnerships intended to support the construction of the NSDI be allowed to focus on that goal.

STATEMENT OF TASK

The Mapping Science Committee will assess the success and potential of the various partnership programs for geospatial capabilities, and how these and future programs based on them contribute to the goals of the broader National Spatial Data Infrastructure. Specifically, the committee will assess the success of the partnership programs in:

- reducing redundancy in geospatial data creation and maintenance,
- reducing the costs of geospatial data creation and maintenance,
- improving access to geospatial data,
- improving the accuracy of geospatial data used by the broader community.

The study will use the status quo in the absence of these programs as the baseline. The study will specifically avoid comment on any additional objectives of these programs that are outside the immediate domain of NSDI.

The success of future partnership programs should be assessed by determining, in a rigorous fashion, how these NSDI partnerships have reduced redundancy in geospatial data collection and maintenance; reduced overall costs in performing these tasks; improved access to geospatial data; and improved the accuracy of the data used. Because much of the FGDC's effort has been devoted to promotion of the NSDI, there has been little opportunity to develop programs that can monitor long-term effects. The FGDC should develop metrics that can be used to monitor long-term progress in the adoption of the principles and programs of the NSDI among agencies at all levels of government, academe, and the private sector. In addition, funding should be directed to projects that are of a sufficient scale to provide well-designed empirical tests of the hypotheses underlying the NSDI goals, and should allow for adequate documentation and dissemination of results.

We found that the programs funded through the FGDC provided only a minuscule proportion of the total resources available nationally to support geospatial data partnerships. It may be that the critical evidence required to demonstrate reductions in redundancy and costs, as well as improvements in accuracy, already exists for partnerships that have developed independently of the FGDC programs. The committee recommends that future partnership programs initiated by the FGDC should be conceived in the context of all relevant partnership programs, and should be designed to augment and leverage them.

It is clear that the efforts of the FGDC to fund partnership activities may be only one of many ways to further the development of the NSDI. The sense of the committee is that we are at an important point in the evolution toward the ultimate goal envisioned by the Committee. New nationwide spatial data are available from the 2000 decennial Census of Population and Housing. The effort of the Office of Management and Budget's new initiative, Collecting Information in an Information Age, has received considerable attention in the last year. Efforts to develop a new organization, the Geographic Data Alliance, are too early to assess. At the same time, local governments and the private sector are devoting considerable resources to complete spatial data they need to serve business clients and the citizens in their communities. All of these activities suggest that the need for a robust NSDI is more important than ever and that it is appropriate for the MSC to continue to monitor and assess the status of the institutional settings and technical progress that affect the development of a robust NSDI.

1
NSDI and Partnerships

GOALS OF THE NSDI

Our nation must continually address a wide range of complex economic, social, and environmental issues. Geospatial information, together with related computer systems, is pivotal to helping communities, companies, and governments synthesize the information required to address these issues in a timely and efficient manner. However, many of these geospatial data are also difficult to locate, obtain, and integrate—in addition to representing a sizable financial investment by each user sector.

The National Spatial Data Infrastructure (NSDI) was envisioned as a way of enhancing the accessibility, communication, and use of geospatial data to support a wide variety of decisions at all levels of society. The National Research Council (NRC, 1993; p. 16) initially described the NSDI as:

> "the...means to assemble geographic information that describes the arrangement and attributes of features and phenomena on the Earth. The infrastructure includes the materials, technology, and people necessary to acquire, process, store, and distribute such information to meet a wide variety of needs."

The importance of the NSDI subsequently was recognized at the federal level in the 1993 Reinventing Government report. The 1994 Executive Order 12906 supported implementation of the NSDI, and the task of providing leadership of the NSDI was assigned to the Federal Geographic Data Committee (FGDC). The importance of the NSDI was reiterated in a 1998 report, *Geographic Information for the 21st Century: Building a Strategy for the Nation* (NAPA, 1998), in which the National Academy of Public Administration identified the NSDI as an important national priority for the United States.

At the core of the NSDI is the concept of partnerships, or collaborations, among different agencies, corporations, institutions, and levels of government. Partnerships are designed to share the costs of creation and maintenance of geospatial data, in order to avoid unnecessary duplication, and to make it possible for data collected by one agency at a high level of spatial detail to be used by another agency in more generalized form. The concept of NSDI partnerships specifically does not refer to joint data ownership, but rather emphasizes the mutual advantages arising from collaboration between partners. Partnerships provide a mechanism for augmenting a system of centralized production of geospatial data, where one (usually national) agency has assumed all of the responsibility and cost, so that the data may be disseminated through a coordinated but diverse patchwork of arrangements that is more suited to meeting local needs. The concept was elaborated in a 1994 NRC report, *Promoting the National Spatial Data Infrastructure through Partnerships* (NRC, 1994), which suggests that given a network of partnerships and effective coordination among partners, the NSDI has enormous potential to minimize the redundant collection of spatial data, to increase citizen participation in decision making, to improve information available to support decision making at all levels of government and the private sector, and generally to sustain the economic well-being of the nation.

Considerable progress has been made in the evolution of the NSDI in the seven years since 1994. For example, the Open GIS Consortium (OGC, 2001), a not-for-profit organization with more than 200 corporate, agency, and institutional members, has made much progress in overcoming the lack of interoperability between geospatial datasets and software systems. Of particular interest is the Web Mapping Testbed, which demonstrates that diverse datasets residing on distributed servers can be combined into a common view through a simple browser interface. Many partnerships have been formed, often at the instigation or with the financial support of federal programs. These partnerships have taken many different forms with many different sets of objectives. The NSDI continues to expand and to reach into new areas of application.

Figure 1 summarizes the current level of involvement in NSDI throughout the nation, based on responses to a recent survey carried out by the National States Geographic Information Council (NSGIC) and the FGDC. Even though there has been considerable success, several questions remain concerning the original premise of the NSDI and the role of partnerships:

- What forms of partnership work best?
- How effective are partnerships at fostering each of the basic aims of the NSDI?
- How successful have the various partnership programs sponsored by the federal government been at achieving the objectives of the NSDI?

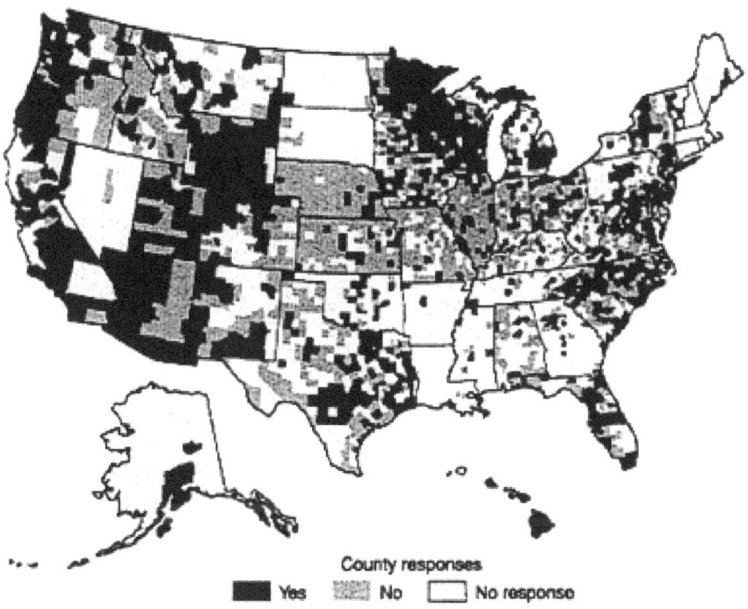

FIGURE 1. Interest in the NSDI is widespread, a result in part of the partnership programs sponsored by the FGDC. This map illustrates how counties responded to a recent NSGIC/FGDC survey asking about active participation in NSDI framework development (Somers, 1999, page 9; reprinted with permission from Geospatial Solutions).

In accordance with the committee's charge to provide external advice to federal agencies, this study is directed specifically to the third of these three questions (see Box 1). More precisely, it addresses the effectiveness of the FGDC partnership programs at meeting the four main goals of the NSDI:

- Reducing redundancy in geospatial data creation and maintenance.
- Reducing the costs of geospatial data creation and maintenance.
- Improving access to geospatial data.
- Improving the accuracy of geospatial data used by the broader community.

If it can be demonstrated and publicized that partnerships are an effective mechanism for achieving these four goals, then the NSDI can be expected to continue to grow and flourish. The committee believes that success in each of these four areas is crucial for the long-term growth and viability of the NSDI.

BOX 1 STATEMENT OF TASK

The Mapping Science Committee will assess the success and potential of the various partnership programs for geospatial capabilities, and how these and future programs based on them contribute to the goals of the broader National Spatial Data Infrastructure. Specifically, the committee will assess the success of the partnership programs in:

- reducing redundancy in geospatial data creation and maintenance,
- reducing the costs of geospatial data creation and maintenance,
- improving access to geospatial data,
- improving the accuracy of geospatial data used by the broader community.

The study will use the status quo in the absence of these programs as the baseline. The study will specifically avoid comment on any additional objectives of these programs that are outside the immediate domain of NSDI.

COORDINATION AND LEADERSHIP

The FGDC was formed in 1990 through the revision of Circular A-16 of the Office of Management and Budget to "...promote the coordinated development, use, sharing, and dissemination of surveying, mapping, and related spatial data" (OMB, 1990; p. 5). The major objective of Circular A-16 was to encourage agencies to avoid duplication of data acquisition efforts. Better data management should minimize the total costs in mapping and spatial data activities, while maximizing the availability of data to large numbers of users. In 1994, Executive Order 12906 directed the FGDC, within the context of the NSDI, to foster coordination among federal agencies, to assist in the development and promulgation of standards, to assist in the identification of requirements for and approaches to producing data, to help develop better means to find and access data, to promote education and training activities, and to facilitate and foster partnerships and alliances within and among various sectors to accomplish all of these activities (Federal Register, April 13, 1994; p. 17671–17674).

At the time of the 1994 Executive Order, the NSDI was still an unfamiliar concept to many in the geospatial data community. The appropriate roles of all levels of government and the various private sector companies were poorly defined, and the steps needed to redefine traditional roles in the NSDI era were not clear. The infrastructure often appeared chaotic with no coherent direction. Organizations were confronted by myriad problems, confusing policies, and even disincentives to coordinate their activities. In addition, many of the essential components necessary for the NSDI to flourish were in their infancy. Soon after the Executive Order, the FGDC made significant advances by effectively communicating the NSDI challenge through newsletters, magazines, and professional journals, and through the organization of national forums. Although geographic information councils had already been formed in many states, the FGDC encouraged their formation in all states, gave its support to the National States Geographic Information Council (NSGIC), and gave that organization a role in FGDC's deliberations. This increased awareness of the NSDI in the geospatial data community and the need for broad-scale coordination to meet NSDI objectives.

A report by the National Academy of Public Administration (NAPA, 1998) drew attention to the need for a statutory basis for the NSDI in order to continue the advances achieved by the FGDC. It recommended the restructuring and consolidation of basic geographic information functions into a new Geographic Data Service, and the creation of a new private, non-profit organization, the National Spatial Data Council, to complement the federal functions of the Federal Geographic Data Committee (Box 2). To date, no formal actions have been taken to implement the NAPA proposals.

BOX 2 STATEMENT BY THE NATIONAL ACADEMY OF PUBLIC ADMINISTRATION

The panel believes that legislation is needed (to sustain the NSDI), but the case for any measure beyond the current executive order still needs to be made. Such a statute, at minimum, should include:

- a list of congressional findings about GI [geographic information];
- a statement of national goals and a definition for NSDI; a charter for the National Spatial Data Council (see below);
- orders for the consolidation of federal base GI functions;
- modifications to existing law to facilitate GI partnerships, cooperative research and development agreements (CRADAs), and private- sector procurements;
- amendments or rescissions of current law to modernize and conform existing program authorizations to the NSDI concept.
  Recommendation:

- Draft a new statute in cooperation with state and local governments and other organizations to create an NSDI, establish a National Spatial Data Council, and better define federal agency roles and responsibilities for NSDI so as to meet the participating organizations' programmatic needs.
  SOURCE: NAPA, 1998; Page xiii.

In addition to the efforts of the inter-agency FGDC, many individual agencies have made concerted efforts to address the need for geospatial data integration. For instance, in 1999 the U.S. Geological Survey (USGS) established the position of Geographic Information Officer, with a mandate to coordinate geospatial data production, maintenance, and integration across the agency, and to build a more integrated interface between the agency and the users of its services.

COMPONENTS OF THE NSDI

The NSDI has been implemented by defining and promoting data and metadata standards, and by establishing a distributed geospatial data 'clearinghouse,' within the context of an overarching data Framework:

Data Standards

Two major standards have been developed over the past decade as components of the NSDI. The Spatial Data Transfer Standard (SDTS) defines terminology and content for geographic datasets. It has been mandated as Federal Information Processing Standard (FIPS) 173 (NIST, 1994), and several federal agencies have developed customized versions of the SDTS to meet their specific needs. While this standard has been mandated for federal activities, its use outside the federal government is essentially voluntary. In the private sector, it competes with a range of standards and formats, many of many of which are associated with specific commercial software products. Moreover, the SDTS competes with other standards in use by the military. In practice, therefore, the general community's adoption of format standards is driven at least in part by the popularity of software products, and time-consuming and expensive conversion between different formats is still common. Although vendor-specific formats may be more popular than SDTS in practice, it must be acknowledged that the effort to develop SDTS provided an opportunity for the community to openly discuss and develop some consensus about the need and mechanism for sharing data between proprietary systems. In some ways, the process leading to the development of this standard was a predecessor to the Open GIS Consortium.

Similarly, the FGDC has done a remarkable job of developing a wide range of standards for the capture, coding, definition, storage, and transfer of spatial data. One of the most important has been the Content Standards for Digital Geospatial Metadata (CSDGM, 2001) that establish the standardized description of geospatial datasets. In part because of the importance of effective description to data sharing and the avoidance of duplication, this metadata standard has had a much more significant effect on the NSDI than the data transfer standard. It has been widely adopted in the geospatial data community within the United States, and it represents the de facto standard around the world. Many other metadata standards are sufficiently similar to the CSDGM that conversion between them is straightforward and supported by software tools. In addition to the six SDTS and metadata standards, the various working groups of the FGDC have now endorsed another 10 content standards for themes such as wetlands, utilities, soils, and vegetation. They have also provided standards for orthophotography, Global Positioning System (GPS) data, and remote sensing. Approximately another 20 standards are in various stages of development. The promulgation of these spatial data standards represents an extraordinary effort by a huge number of agencies and individuals. The FDGC should be applauded for encouraging and facilitating these developments.

National Geospatial Data Clearinghouse

Over the past seven years, the establishment of the National Geospatial Data Clearinghouse (NGDC, 2001) has emerged as an important operational component of NSDI. This web-based data server technology represents an excellent example of how the FGDC has reacted to the 1994 Executive Order. It consists of a small number of portals, or access points on the Internet, that provide identical services, together with a larger number of servers that provide direct access to geospatial datasets. The data clearinghouse appears to users as a single, virtual, geospatial data catalog. Portals and servers are maintained by their sponsors, which include federal and state agencies, non-governmental organizations (NGOs), universities, and corporations. Each server's sponsors contribute data and associated metadata records—using the metadata standards—and manage both data and metadata records locally. At the time of writing there were six portals, more than 250 servers in 26 different countries, and several thousand datasets in the clearinghouse system. As with the metadata standard, the FGDC has taken a lead role in the implementation of standard web-based data serving. The clearinghouse standard has proven very popular with both its sponsors and its users, and has become the de facto international standard. Even though other internet-based solutions for distributing spatial data have evolved in both the public and private sectors, the pioneering effort of the FGDC to demonstrate the feasibility of the concept must be acknowledged.

NSDI Framework

The core of the NSDI is data sharing, and accurate data must be constructed on a solid foundation. Although a very large number of geospatial data types exist, those that constitute the critical base layers are considered to be the framework for the entire system. The MSC's 1995 report, *A Data Foundation for the National Spatial Data Infrastructure,* articulated the need for a NSDI foundation. The committee used the construction of a building as a metaphor: "...A solid foundation of concrete or other material is first put in place; then a framework of steel beams is connected to the foundation to create a structure to support the building's interior and exterior" (NRC, 1995; p. 15). In the same way, a foundation of spatial data serves as a reference for integrating other data themes. As these themes are developed and integrated with the foundation, a structure will be created that can support and sustain the NSDI. The committee considers geodetic control, digital terrain, and digital orthorectified imagery to constitute the NSDI foundation. Under Executive Order 12906, the FGDC established subcommittees and placed priority on transportation, boundary, and hydrology data. In 1995, the FGDC framework working group identified the purpose and goals for the framework and incentives for participation; defined the information content; developed preliminary technical, operational, and business contexts; specified the institutional role needed; and developed a strategy for a phased implementation of the framework (FGDC, 1995). The Framework Working Group identified the following seven themes as the framework of the NSDI:

- geodetic control (the measurements and monuments that form the foundation for practical surveying);
- orthoimagery (digital datasets derived from aerial photographs and corrected for distortion);
- elevation (digital files defining the height of the land surface and depths below water surfaces);
- transportation (roads, railways);
- hydrography (rivers, lakes, reservoirs);
- the definition of boundaries and names of government units (counties, states, cities); and
- cadastral information (boundaries defining land ownership).

In a sense, the NSDI Framework is the digital equivalent of the USGS's topographic map. Separate layers representing topography, transportation, hydrology, and cultural features—each denoted by a specific color and cartographic symbol—comprise the topographic map. The topographic map also relies on a solid foundation of geodetic control and imagery. For decades, numerous data acquisition and presentation activities have been based on the USGS topographic quadrangles. Users tie other information to these maps, either through annotation or by directly overlaying information on transparent sheets. As the USGS converted these maps into digital line graphs, the information on these maps became the first nationwide de facto spatial data framework. Over the past couple of decades, other federal agencies, such as the Census, have taken data from these original topographic maps, edited them, and added topological structure and attributes to meet their individual needs.

Arguably, the objective of the NSDI is to improve the spatial resolution, the accuracy, the content, and the currency of this base. As the FGDC (1997a) notes, the NSDI Framework should also consider the procedures and technology for building and using the data; and the institutional relationships and business practices that support those procedures. It is the institutional partnerships that are the focus of this report.

PURPOSE OF THE REPORT

Over the past seven years, since the 1994 Executive Order and the MSC's report, *Promoting the National Spatial Data Infrastructure through Partnerships* (NRC, 1994), the NSDI has matured considerably. The FGDC has made significant efforts to reach local and state governments, both in funding initiatives and in developing policy. Much excellent work has been done in promoting the NSDI's core ideas, developing consistent standards for the representation of spatial data, and raising awareness of its objectives among the broader geospatial data-user community. Such awareness is essential if the NSDI is to succeed, because NSDI is, by definition, a community effort.

Seven years after the Executive Order, the NSDI is moving into the next phase of its institutional development. Only through a concerted effort will the NSDI succeed in its goal of reforming the production, dissemination, and use of geospatial data. Growth to date has been sustained largely by belief in the principles of NSDI, rather than by any hard evidence of success, and the concept of partnerships expounded by the MSC in its 1994 report remains largely an unrealized construct. As we enter the new millennium, the National Research Council considered that it would be valuable for the MSC to assess the success and potential of the various FGDC-sponsored geospatial-data partnership programs, and to assess how these programs and future programs based on them contribute to the goals of the broader NSDI. Specifically, the committee was tasked to assess the success of the partnership programs in: (1) reducing redundancy in geospatial data creation and maintenance; (2) reducing the costs of geospatial data creation and maintenance; (3) improving access to geospatial data; and (4) improving the accuracy of geospatial data used by the broader community (Box 1). In its 1994 report, the committee had argued that all four of these effects would follow from the implementation of partnerships under the umbrella of the NSDI. In a sense this report provides a barometer of whether the FGDC programs are fostering these outcomes.

2
Review of NSDI Partnership Programs

The MSC, in its 1994 report, *Promoting the National Spatial Data Infrastructure through Partnerships,* stated (NRC, 1994; p. 1): "Cooperation and partnerships for spatial data activities among the federal government, state and local governments, and the private sector will be essential for the development of a robust National Spatial Data Infrastructure (NSDI)." In this report, the committee articulated its vision of a partnership model. This model was built on the foundation of shared responsibilities, shared cost, shared benefit, and shared control. That same report reviewed some existing cooperative programs and partnership activities. These included:

- The Bureau of the Census's State Data Program.
- The National Geodetic Survey's program for incorporating local input into the national geodetic control network.
- NOAA's partnership with South Carolina to build a state-of-the-art natural resource information management system.
- EPA's cooperative program to help fund the Maryland Digital Orthophoto Program.

These are just a few examples of how federal agencies work with nonfederal partners to help advance the development of spatial data. Some of these programs may be viewed as mechanisms for meeting agency mandates, whereas others are based on special funding arrangements. Clearly, such partnership activities would have evolved out of necessity, innovation, or political motivation without the existence of the FGDC.

The MSC has always viewed the NSDI in the broadest possible context. It suggested that "The infrastructure includes the material, technology, and the people necessary to acquire, store, and distribute…geographic information that describes the arrangement and attributes of features and phenomena on the Earth" (NRC, 1993; p. 2). In its workshop report, *The Future of Spatial Data and Society* (NRC, 1997; p. 42), the MSC concluded that "The NSDI is comprised of consortia in which all stakeholders in the spatial data community play important roles, whether as federal, state, or local governments; corporations; academic institutions; or individuals." This broad definition makes it extremely difficult to assess the status of the NSDI in its entirety; therefore, this study only focuses on the specific role of the FGDC as a catalyst in the process.

When the FGDC was given the explicit mission of coordinating federal spatial data development activities, it identified the need to establish a more formal mechanism for developing partnerships. This chapter presents a review of the FGDC partnership programs that have promoted various aspects of the NSDI over the past seven years. As a consequence of the MSC's charge to provide external advice to federal agencies, the primary focus of this study is to review these specific FGDC-sponsored programs rather than to assess all the other formal and less formal programs sponsored or coordinated by non-federal groups and institutions that have also helped to promote the development of the NSDI. This review includes a brief discussion of each program and its objective, together with an assessment of the program's effectiveness in addressing the goals of the NSDI. These assessments rely on: views the committee gathered through presentations made at its September 1, 1999, meeting; on past assessments the sponsors of partnership programs conducted and made available to the committee; on views participants expressed in a forum the committee convened at the August 1999, NSGIC meeting in New Orleans; on responses to a questionnaire the committee distributed to participants in federally sponsored partnership programs; and on the committee members' experience and expertise. While it is impossible for the committee to conduct in-depth surveys, its members have extensive firsthand knowledge of the development of the NSDI, the programs of the FGDC, and the experiences of a wide range of users in both the public and private sector.

It should also be noted that in February 2000, the University Consortium for Geographic Information Science (UCGIS) issued a Request for Proposals (UCGIS, 2000a) for an assessment of FGDC's funding programs, "...to better understand their effectiveness, to determine whether the grants are the most effective means to achieve the NSDI goals, and to help guide future grant efforts." The UCGIS study, being carried out by the Department of Geography at SUNY at Buffalo, is being funded under a contract between UCGIS and FGDC. We expect that, when completed, the UCGIS study will add substantially to our knowledge of the effectiveness of these programs, and will complement the content of this present report. An important element of this study is an assessment of reasons why organizations have decided not to participate in the NSDI partnership programs.

NSDI COOPERATIVE AGREEMENTS PROGRAM

In 1994, the FGDC initiated the NSDI Cooperative Agreements Program (CAP) (1994; p. 1) "...to help form partnerships with the non-federal sector that will assist in the evolution of the NSDI. The goal is to encourage resource-sharing projects through the use of technology, networking, and more efficient interagency coordination..." This program is now in its seventh year. It funds activities that promote the goals of NSDI, and is designed to provide relatively small amounts of money that leverage local sources and stimulate new activity, particularly new partnerships. By keeping the funding amounts small and limiting its awards to seed funding for one year, CAP strives to initiate long-term activity while avoiding long-term dependency on federal funding.

As a consequence of policy decisions and budget priorities, the nature and size of the program and the types of projects funded have varied considerably from year to year:

1994	—Approximately $250,000 was distributed among nine projects.
1995	—Projects that developed and used metadata tools were emphasized; $625,000 was allocated to 22 projects.

1996	—$1.1 million supported 31 projects, with an emphasis on Framework development.
1997	—$1.2 million was allocated to support 36 projects, with an emphasis on projects involving many cooperating groups.
1998	—$1 million was given to 31 projects; for the first time federal agencies were eligible for funding.
1999	—$1.8 million was used to support 95 projects under the Don't Duck Metadata program (see below). The funding success rate was very high (95 of 108), but the grants were smaller than in previous years. Before 1999, funding success rates had averaged 35 to 40 percent.
2000	—$1 million was distributed among 45 projects, supporting the Don't Duck Metadata initiative, Framework Demonstration projects (see below), and for the first time the Open GIS Consortium's Web Mapping Testbed was supported with four successful projects.
2001	—$1 million is available for partnership projects, distributed among Don't Duck Metadata, integration of clearinghouse nodes with the Web Mapping Testbed, and U.S.-Canadian Framework collaborative projects.

To be eligible for CAP funding, a proposal must involve a partnership among agencies, and non-federal partners must provide matching funding of at least 25 percent of total project costs. Successful CAP projects have usually included an emphasis on improvement of local government decision making. The funds have been used to encourage new partnerships that can build on existing expertise. They have typically addressed one or more of three fundamental areas of data sharing: improving the way users find or access data through the development of clearinghouses; improving the integrity or usability of data through the creation of metadata; and creating or maintaining the data themselves. A few projects have simply promoted the concepts of the NSDI or provided training and educational opportunities.

Several practical problems have arisen in the management of CAP funding. Since the grant competition is based on an annual cycle, some states with a biennial budget process have not been able to respond in a timely manner. Furthermore, because the grants are fairly small, institutional oversight has not always been adequate. For example, a few grants have been awarded to smaller agencies that are not part of a statewide coordinated effort. In fact, some awards have detracted from long-term state objectives by diverting resources from data conversion efforts. The FGDC has resisted requiring state geographic information councils to approve proposals, but does look for consistency with state strategic plans. It also favors proposals that appear to promote attention to NSDI issues at the state level; however, the FGDC has no formal mechanism to ensure that the funds are compatible with local goals.

Over the past seven years, every state except North Dakota has presented a successful proposal. Although the committee has not conducted a comprehensive analysis of the success and impact of all of these awards, anecdotal evidence suggests that some states have certainly utilized the CAP funds to assist and promote ongoing efforts. States that have been most successful in gaining awards tend also to be most actively involved in other aspects of the NSDI. For example, they are likely to have established a state geographic information council, developed a clearinghouse node, or had a high response rate to the 1998 NSGIC Framework Data Survey. The ultimate success of the NSDI will depend on nationwide acceptance. While it is unlikely that each of the thousands of local government entities will endorse the objectives of sharing spatial data, it would be reasonable to expect every state to participate. The FGDC, in conjunction with NSGIC, could establish a virtual organization (an "Interactive Town Square," see OMB, 2000) to keep everyone informed and make organizations aware of opportunities. Efforts could be made through the National Governors Organization to designate a key office in a state that would be charged with the responsibility of handling communication with the FGDC. The FGDC could also concentrate on educational or training sessions that could be offered regionally. This would encourage regional participation and minimize the cost to participants.

In September 1997 the FGDC produced its own assessment of the CAP, based on the period 1994 through 1996 (FGDC, 1997b). The FGDC report examined program effect from three perspectives:

- *Program output:* were project objectives achieved?
- *Intermediate outcome:* are project efforts being continued beyond the funding period?

- *Long-term impact:* are the tenets of the NSDI being incorporated into the programs of non-federal organizations?

The FGDC's information was obtained from final project reports as well as from questionnaires sent to the 62 funding recipients, 52 of whom responded. The FGDC assessment concluded that the CAP is:

- adding structure and discipline to the process of building a national information resource;
- helping state governments, libraries, universities, local government organizations, and private sector entities become anchor tenants on the NSDI and thereby attracting others to use and become a part of the infrastructure;
- helping to form data-sharing partnerships that are still continuing, that might otherwise not have happened;
- increasing the level of collaboration across agencies, and bringing attention to organizations that has led to new collaborative activities;
- showing the non-federal community the importance of documenting data to standards that will make the data useful in multiple applications;
- raising the level of information technology skills in the geospatial data user community as project collaborators train people in their local communities, who in turn become trainers of others;
- building the accumulation of experience and knowledge that others can use to reduce the uncertainties associated with investing in new ideas and technologies and, ultimately, lower their costs;
- showing the non-federal sector the feasibility of some applications that they might otherwise have passed over;
- changing, in some cases, agencies that have historically been information repositories to being customer-driven service providers;
- extending access to the NSDI to new constituencies and to organizations and communities that typically are not on 'the geospatial information highway'; and
- clearly demonstrating that as completed projects have time to mature and grow, organizations are realizing more benefits than originally anticipated.

In the committee's opinion, the total financial commitment to the CAP program represents a very minor investment. The total federal contribution to the CAP during these three years was approximately $2 million. By comparison, the Office of Management and Budget estimated that total federal expenditures on digital geospatial data activities in 1993 amounted to approximately $4 billion, and total sales of GIS software in these years were in the hundreds of millions. To emphasize this point, a recent commentary estimated the total worldwide expenditure on GIS and related activities was of the order of $15 billion to $20 billion (Longley et al., 2001; p. 360). An examination of personnel costs provides a useful perspective on the CAP investment. The $2 million investment would provide full-time employment for at most 20 suitably trained people for one year. That averages approximately half a person-year for each of the states that were successful in the program. Even under the most optimistic leverage scenarios, CAP funding was only a minor component of total geospatial data investment. It is to the FGDC's credit that CAP recipients are so positive about the experience, and the program has seeded so many projects that have the potential for long-term effect. This is particularly noteworthy given obvious constraints the one-year budget cycle imposes on these projects.

Framework Demonstration Projects Program

The Framework Demonstration Projects Program (FDPP) was initiated in 1996 as a funding initiative separate from the CAP. In 1998 a joint announcement of both programs was made, and in 1999 the program was merged with the CAP program. Continued support for the FDPP was reflected by the funding of four projects in 2000, but the program was not included in the 2001 call for proposals. According to the FGDC (1996; p. 1), the FDPP was established to:

> "...support cooperative projects that test the means by which the geospatial data community can work together to build and maintain the data Framework for the NSDI...Funding is provided for implementations of multi-organization, multi-sector partnerships to coordinate data collection, maintenance, use and

access in local and regional areas. Program participants will identify a basic information content for the Framework data and will develop technical, operational, and business contexts by which a distributed, collaborative data collection and maintenance effort will operate."

At approximately $100,000, the average FDPP award is substantially larger than the average CAP award made between 1994 and 1998, and almost an order of magnitude larger than the CAP awards of 1999. In 1996 the FDPP funded seven projects for a total of $810,000. Total funding for the program fell to $470,000 in 1997, but rebounded to $845,000 in 1998. The following examples illustrate the range of projects funded under the program:

- *A Statewide Framework of Public Lands Data Using Locally Derived Cadastres* (North Carolina, 1996) "...will create a viable technical process for the maintenance of the Framework cadastral theme in North Carolina by improving statewide datasets of federally and state-owned property" (FGDC, 1997a; Appendix H).
- *The Baltimore-Washington Regional Digital Spatial Data Framework Demonstration Project for the Gwynns Falls Subwatershed* (Maryland, 1996) "...will explore the administrative and technical issues of linking local and regional datasets for the Framework themes of geodetic control, digital orthoimagery, elevation, transportation, hydrography, governmental units, and cadastral data" (FGDC, 1997a; Appendix H).
- *Alaska Transportation Mapping Coordination Project Linking State and Local Programs to Build the NSDI* (Alaska, 1998) "...to better organize Alaska's state and local mapping authorities to address the transportation Framework layer" (FGDC, 1998; p. 1).

Even though the funding level for these projects was more substantial than in the previous FGDC effort, the amounts remain small in comparison with the size of the geospatial data user community. Furthermore, the committee finds it difficult to determine whether the larger FDPP grants have been more effective than the smaller CAP grants, although it is apparent that the relatively small size of CAP awards and their short duration has created some problems of continuity. The core team for all of the projects tends to be small, and the temporary nature of the funding often leads to an unstable working environment. The departure of a key player can severely impede the success of a project and momentum can quickly disappear.

Don't Duck Metadata

In 1999 the FGDC sponsored 95 projects to promote the creation and use of metadata in support of geospatial data sharing. This program was designed to encourage the adoption of consistent policies for metadata, and to counter the notion that metadata are expensive to create and have limited benefits. Grants of approximately $18,000 were given to 42 states to stimulate partnerships that would promote the development of metadata.

Metadata play a critical role in the NSDI. They facilitate the sharing of data, particularly between partners who are not in direct contact with one another; it is necessary to document the contents of datasets; to provide sufficient detail to allow computing systems to open and access them; and to document data quality. In effect, these metadata components allow potential users to assess the fitness of datasets for their own use, and to minimize the problems associated with importing data from another system.

Such sharing of data is central to the NSDI goals of reducing duplication of effort, improving data quality, and improving data access. Unfortunately the benefits and costs of metadata creation accrue in ways that do not necessarily promote these goals. Most of the costs of metadata creation accrue to the custodians and creators of data, while most of the benefits accrue to users, often in other organizations. As a result, data providers tend to "duck" metadata or to assign them a low priority. The FGDC believes that one solution to this difficulty is to bring users and creators into a single partnership that can reassign or aggregate costs and benefits in ways that are more satisfactory to all the partners.

The committee considers that smaller grants (e.g., the average award of $18,000 in 1999, and $22,200 in 2000) appear to be inadequate to meet the program's objectives. Moreover, the decision to fund almost all applicants (95 of 108 in 1999; 31 of 32 in 2000) may prove to be detrimental to the goal of partnership development. After reviewing the call for proposals for the 2001 program, the committee notes that the maximum grant limit—$6,000 for metadata implementation assistance and $30,000 for trainer assistance—may restrict the likelihood of success, whereas the maximum limit of $75,000 for the single US-Canadian Framework project is a more appropriate funding level. The committee also noted that the 30 percent of the funding allocation for 2001 is reserved for federal agency grants. The committee considers that the strength of the NSDI partnership program lies in the development of partnerships between federal agencies and other levels of government, industry, and academic communities, and views the reservation of such a substantial proportion of available funds as detrimental to the leveraging concept and unlikely to have the catalyzing effect that the committee originally promoted. In the committee's view, one of the significant benefits of the FGDC partnership programs lies in the effort that must be made during the proposal preparation stage. Several participants have commented that the "carrots" the FGDC offers have fostered interagency cooperation, which has resulted in successful long-term collaborations independent of the outcome of the award competition. Consequently, a high success rate may actually reduce one of the incentives for collaborative efforts.

COMMUNITY DEMONSTRATION PROJECTS

Community Demonstration Projects represent an FGDC effort to promote another level of partnership. By using $645,000 provided by the National Performance Review Fund, FGDC was able to fund six projects from 1998 to 2000 that demonstrate the importance of geospatial data in community-wide planning. At its September 1999 meeting, the MSC heard presentations on the program as a whole, and on projects carried out in Dane County, Wisconsin; the Tijuana River Watershed; Gallatin County, Montana; and the Baltimore City Police Department. Environmental Systems Research Institute, Inc. (ESRI) is a partner with FGDC in these activities, providing the projects with in-kind software support and expertise.

Each of the projects is exploring an innovative form of community-federal partnership with a major geospatial data component.

A common objective for each of the projects is to promote the broad-based participation of stakeholders in planning and decision making by enabling geospatial information to be easier to create, share, and use. The underlying premise is that geographic information technology can change the traditional way that local decisions are made. The goal is to better inform citizens, to get them involved in the planning process, and to enable them to explore alternative scenarios. Each of the projects makes use of GIS technology, as well as a wide range of alternatives such as the Internet, cable TV, displays at public meetings, and collaborative software to help disseminate information. The following bullets summarize the key points made to the committee on each of the four projects presented:

- *Dane County, Wisconsin.* This county has a long history of innovative uses of geospatial data and technologies, particularly in agriculture and land-use planning. The project aims to address inequities in the accessibility of geospatial information through a series of workshops for professionals and the general public. Among many benefits of the process to its stakeholders, the Natural Resources Conservation Service (NRCS) will gain a better understanding of community needs for soils data. Available at: http://www.lic.wisc.edu/shapingdane.
- *Tijuana River Watershed.* This watershed straddles the United States-Mexico border and feeds an estuary that is part of NOAA's estuarine research program, the National Estuarine Research Reserve (NERR) Program. Geospatial information provides the common language among five overlapping projects within the NERR, with an improved assessment of flood vulnerability as a major goal. Available at: http://typhoon.sdsu.edu/tjwater.
- *Gallatin County, Montana.* This project aims to engage the community in evaluating options for growth in the county, which is being impacted by urban sprawl. The county contains part of the Greater Yellowstone area, with its high environmental sensitivity. This project is exploring and evaluating ways of presenting geographic information and planning options through community meetings, the media, and other mechanisms. Available at: http://co.gallatin.mt.us/planning/index.htm.

- *Baltimore City Police Department.* Conducted in partnership with the Department of Justice, this project is exploring ways to use maps and other products of geospatial information technologies to reduce crime, and the fear of crime, in Baltimore neighborhoods. Available at: http://usdoj.gov/criminal/gis/rcagishome.htm.

Two additional projects were completed under the Community Demonstration Program:

- *Tillamook County, Oregon.* This flood mitigation and restoration project integrated several datasets to assess the risk of flooding. Available at: http://gisweb.co.tillamook.or.us.
- *Upper Susquehanna-Lackawanna Watershed.* This flood mitigation and environmental management project utilized web-based approaches to share maps with the community. Available: http://www.fgdc.gov/nsdi/docs/cdp/ppt/UpperSusquehanna_files/frame.htm.

Although each of these projects revolves around geographic information technology and geospatial data, their objectives and the objectives of their sponsors go well beyond the immediate aims of the NSDI, especially in the realm of data integration. Although the federal investment is small, it is being spent in just a few local areas, and thus can be expected to have a more significant local impact than the same amount of investment directed at an entire state. The committee believes that these projects represent a valuable investment in a few well-designed experiments. The four projects represent a cross section of geographic areas and public policy issues. They are being conducted by experienced teams and show great promise in evaluating how well some of the technological advancements that have progressed and whether they are useful in promoting a broader base of citizen participation. The projects also offer a good opportunity to determine whether the geospatial data are appropriate for the level and type of policy decisions being explored. The FGDC's final report noted that funding precipitated the formation and maintenance of partnerships that would otherwise probably not have developed (FGDC, 2001). The FGDC document, *Overview of the Lessons Learned from the NSDI Community Demonstration Projects* (Executive Summary), emphasizes the importance of vision, capacity, and support. The focus of the projects was on "the capacity to acquire, deliver and use geospatial data and tools in a decision making process." The report also suggests that:

> "...Federal grant dollars can provide an effective incentive for communities to embrace NSDI standards and serve as "seed money" for purposes of leveraging financial and technical resources from other sources.... The NSDI community should initiate and expand projects to initiate a national infrastructure that focused on community data and information needs and eliminates barriers that communities face in working with the Federal government to build place based information management systems."

It is important to note that in June 2000, the National Partnership for Reinventing Government (NPR) gave a Hammer Award to the NSDI Community Demonstration Projects. The NPR is an interagency task force established in 1993 to find ways to make government "...work better, cost less and get results Americans care about..." (NPR, 1993). This award recognizes exceptional achievement in reinventing government. The Community Demonstration Projects were recognized because they show the benefits that can be realized by an expanded sharing of geographic information among federal and local agencies. While the Hammer Awards may not represent a very rigorous evaluation of the merits of these projects, the committee believes that it is significant that NSDI-oriented projects supported by the FGDC are being recognized as important ways to make government more cost effective and efficient.

COMMUNITY-FEDERAL INFORMATION PARTNERSHIPS

Development of the Community-Federal Information Partnerships (CFIP) concept was proposed in 1998 as an initiative involving several federal agencies; it evolved into a $20 million proposal in new federal funding in FY 2001. Its goals are similar to those of the Community Demonstration Projects, but encompass a much broader domain and with a much higher level of federal investment. The CFIP program focuses on the role of geospatial information in community planning and the development of "livable communities"; on the role of the federal government as an agent of change and as a coordinator of geospatial data infrastructure; and on the ways that data, metadata, and technologies can be deployed to make geospatial data more accessible to all of the community's stakeholders. A major goal of the program would be to demonstrate that the NSDI is the key to integration, and that it constitutes a way of coordinating federal and local interest in solving local issues. CFIP has received strong support from the NSGIC and from the National Association of Counties (NACo).

In the committee's view, the proposed CFIP program will have to resolve several issues in order to be successful. An analysis of project scale is needed to clarify what can be achieved with any specific level of funding, or how to divide the total funding among projects to achieve maximum effect. The committee suggests that careful consideration be given to whether the program's objectives might be better served by a few large grants, as in general it considers that a much larger number of small grants may not always be effective. The committee advocates adoption of a funding formula that provides resources to all participants on a non-competitive basis, coupled with grants of sufficient size and duration to achieve expected outcomes. As a multi-agency program, the goals of the program are very diverse, go well beyond those of the NSDI, and will have to be clearly articulated if the program is to be successful. The process by which funds are awarded will have to be clarified, as it will involve multiple agencies and stakeholders at all levels, all ideally working toward common objectives.

PRIMING THE PUMP-THE FEDERAL ROLE IN NSDI PARTNERSHIP INITIATION

By definition, the NSDI is an ambiguous concept. It is not an end in itself, but rather a means to an end. Although it could be argued that spatial data should be treated as a commodity that is created and distributed according to a simple business model, the committee believes that it should be treated as a public good. Ultimately, geospatial data exist to serve societal purposes, such as the mitigation of hazards, efficient operation of delivery services, and wise manage ment of natural resources. Geospatial data are a collective resource, produced and used by many different groups, agencies, and individuals. In this context, the NSDI represents a mechanism for the more effective production, management, and use of geospatial data. It can also be viewed in the context of a substantial innovation in the way that data are traditionally created and managed. Therefore it is useful to examine the motivation, the impediments, and the rate of adoption as this innovation diffuses through society. The NSDI inherently falls within the larger domain of information technology; therefore, it is also useful to view its development in terms of whether the intended user community is passing through the set of societal and technological "gates" that Mayo (1985) suggests inhibit the adoption of any new technology. Tulloch (1999) provides an excellent discussion of how the implementation of a multipurpose land information system can be viewed in terms of the conflict between what Mayo describes as the push of technology and the pull of society. The committee envisions the role of the FGDC as an agent of change that is charged with the mission of pushing and pulling a vast and unorganized set of users through these gates. In this sense, the development of partnerships represents successful and demonstrable evidence that the goals of NSDI have been accepted and that diffusion is occurring.

More specifically, the designers of the NSDI argued that its construction would provide four benefits: reduced redundancy in geospatial data production; reduced cost; greater access to geospatial data; and greater accuracy of geospatial data. All of these four imply a comparison between a world with the NSDI and a world without it, or the world that existed before the NSDI was established compared with the situation that would have existed now had the NSDI not been constructed.

In the committee's view, the NSDI is explicitly a national concept in which the federal government originated and continues to play the major role in its construction. This is an appropriate responsibility for the federal government for several reasons. First, there is a natural tendency to equate nationwide and federal, in part because the federal government is the sole government of the nation as a whole, and in part because of its sheer size. Second, and more specifically, the federal government, through the FGDC, has played a major role in the definition and design of the NSDI: The Executive Order that initiated the NSDI may be viewed as an order to the federal government, rather than as an order to the nation. In addition, the federal government clearly has much to gain from the NSDI, as well as deserving credit for much of the work behind its construction. The NSDI can provide much of the geospatial data that the federal government needs to carry out its own programs. For example, the Bureau of the Census depends on local governments for current listings of streets and addresses. A mature and efficient local-federal partnership that successfully overcomes both the technical and institutional barriers that inhibit the sharing of this information could greatly reduce the cost of conducting the decennial census.

Ideally, in a robust NSDI, these data would be continually updated on a transaction basis at the local level, and shared dynamically over an internet-based clearinghouse with federal and private users. Such a partnership would probably result in a more efficient local emergency 911 system and facilitate commercial package delivery services, while simultaneously assisting the creation of a nationwide street centerline database. The committee also notes that the Ground Transportation Subcommittee of the FGDC and the Cultural and Demographic Subcommittee have made considerable progress in developing standards for handling transportation features and street addresses. These draft standards—now out for public review—are the result of extensive review by participants from many agencies. Broad acceptance of these standards will play a significant role in enabling organizations to share street and other transportation data. Because the production of geospatial data is mandated for many federal agencies, including the USGS and NOAA, it is in their self-interest to promote a robust NSDI. In fact, as the demand for higher resolution and more precise spatial data intensifies, it could be argued that the federal mapping functions will become increasingly dependent on local government data sources. For example, recent changes in policy have significantly improved the accuracy of mapping-grade GPS receivers to approximately 10 meters, which is less than the stated accuracy of the 1:24,000 scale USGS topographic quadrangles. This suggests that the largest scale nationwide mapping series is an inappropriate base map for many applications. A serious question for the next decade will be to determine the most appropriate approach to the development of new national map series at a much higher level of accuracy. Although the orthophoto base of the nationwide framework database is based on a 1-meter pixel size, many local governments have already invested in orthophotos with a 0.5-foot resolution. The federal government must find more innovative ways to incorporate these high quality data sources into their overall strategy. It is clear that we are in a period of rapid change in terms of human-computer interaction and institutional arrangements. It is important that the federal government actively monitor the technological setting for the use of spatial data and participates in the further enhancement of applications.

For the benefits of a robust NSDI to accrue, however, it must first reach a threshold of sustainability. The community of geospatial data producers and users must be made aware of its concepts and design, and must be persuaded to adopt them (i.e., pushed through the social gates). Because of its patchwork nature, the NSDI cannot be successful unless a large proportion of the geospatial data community adopts the NSDI principles. According to Rogers (1995), the diffusion of an innovation generally passes through five stages: knowledge, persuasion, decision, implementation, and confirmation. As in any adoption cycle, organizations will vary greatly in the rate in which they progress through these stages. Some innovative groups will have the organizational structure and the technical ability to be early adopters, whereas others face severe impediments that will force them to lag considerably behind. The following is a simple model describing that adoption process, in three stages:

1. Awareness or knowledge of the NSDI is promoted through the efforts of the FGDC, other federal agencies, professional organizations such as NSGIC, and individual advocates. Efforts are made to ensure that local and regional governments are provided with concrete examples of how the use of spatial data can help them solve critical problems. Benefits are characterized as incentives to capture the attention of the community, and additional monetary incentives are provided. Other parts of the community are persuaded by the novelty of the concept, and see benefits in being perceived as trendsetters.

2. Initial adopters make decisions to assist with implementation of the NSDI. They realize its benefits, document them, and disseminate awareness of these benefits to new parts of the community.
3. Residual sections of the community are convinced by the demonstrated benefits of the NSDI, and their actions complete the adoption process.

In the context of this simplified adoption model, the partnership programs discussed in Chapter 2 that are designed to provide financial incentives can be assigned to the first stage. In the committee's view, the FGDC has played an important role in this first phase of adoption. It believes that awareness of the NSDI goals is now widespread among the user community, and there is considerable knowledge of the availability of partnership funding. However, the recent University of Kentucky study conducted for the FGDC (Harvey, 2001) noted: "More surprising, our survey revealed that half the respondents did not know what NSDI referred to. The limited awareness among local governments suggests that the most significant hurdle for developing the NSDI is raising awareness and educating local governments." The study also found that local governments realize that they could benefit from the use of federal data sources but the smaller ones face major obstacles in the adoption of new technology and they feel excluded from the process. The authors conclude (Harvey, 2001, p. 40): "Lacking specific policy, financial, or organizational guidelines to promote involvement, NSDI implementation stumbles at the local level." It must also be noted that although many larger local governments (e.g., Cook County, Illinois, which is investing $15,000,000) have a clear business plan for the use of spatial data, it is not clear that they feel the need to share that data with other levels of government. Continued success of the adoption process will depend on persuading a large proportion of the user community to adopt the design of the NSDI. Harvey suggests that "Building the NSDI is not only a matter of building a pyramid of data, but also of creating a pyramid of trust." The ultimate success of such widespread adoption will depend upon proof of benefits. If that proof does not materialize, the adoption process may terminate at the first stage and never provide the benefits that were originally hypothesized.

As noted in Chapter 1, the committee's purpose in initiating this assessment was to determine whether programs conducted to date have assisted in meeting the four main goals of the NSDI. As dominant sponsors of a first stage of adoption, the federal government has successfully "primed the NSDI pump." This priming action appears to have been primarily directed at the one specific goal of improved access to data, and the evidence the committee gathered clearly demonstrates that the NSDI does indeed improve access to data. The actions of the federal sponsors of the NSDI, in creating the National Geospatial Data Clearinghouse (NGDC) and fostering the use of the Content Standards for Digital Geospatial Metadata (CSDGM) through partnership programs, have led to a substantial improvement in nationwide access to geospatial data. Furthermore, we anticipate that a second stage of adoption will follow; namely, where many more agencies and organizations can be expected to participate in the NGDC and adopt the metadata standard, without requiring further direct pump-priming and encouragement by the federal government. It should also be noted that the FGDC and UCGIS funded the Spatial Data and Visualization Center at the University of Wyoming to develop educational materials on metadata; see http://www.sdvc.uwyo.edu/metadata.educational.html.

THE FUTURE FEDERAL ROLE IN DEVELOPING THE NSDI

Full adoption of the NSDI will require attention to the remaining three goals: reduced redundancy, decreased cost, and increased accuracy. To date, the federal government's funding incentives through the NSDI partnership programs do not appear to have had a significant effect on these goals. In many ways, these additional goals rely on a much more fundamental level of cooperation between partners than the simple sharing of an agency's existing data. Because these goals are critical to the future evolution of the NSDI, the committee considers that continued evolution of the NSDI is in some jeopardy. Organizations that initially responded positively to the NSDI, attracted by the obvious benefits and financial incentives, may grow bored or disenchanted and withdraw when the novelty wears off and the funding disappears. Others who were drawn by incentives provided by federal partnership programs may withdraw when it becomes clear that those incentives were not intended for the long term.

The committee strongly suggests that, to assure the future of the NSDI, attention be directed at the remaining three goals. Specifically, future partnership programs sponsored by the federal government should be required to provide convincing evidence that adoption of the NSDI's concepts and design results in reductions in redundancy and cost, and increased accuracy. These projects should serve as clear models of the benefits of partnerships and mechanisms for long-term sustainability. To be convincing, such demonstrations should satisfy certain criteria:

Scale. Demonstrations should be large enough to provide unambiguous results, and sufficient resources should be provided to ensure that there is sufficient time for the project to be completed.

Visibility. Demonstrations should be widely visible to the geospatial data community, and sufficient resources should be provided to ensure that results are widely disseminated. This can be in the form of virtual town hall meetings and "cookbooks" that demonstrate clear success stories that should be widely distributed at professional meetings attended by local government officials and workers.

Rigor. Demonstrations should be designed according to appropriate scientific principles, with solid experimental designs that will ensure that the findings can be extended to other areas. This should include efforts to better understand the impediments to successful adoption of the goals of the NSDI.

It will also be important that future funding initiatives be widely advertised, with the criteria for selection clearly stated. Ideally, a panel of experts in the field should evaluate the proposals, with appropriate peer-review.

Partnership is a very general concept that can serve many different ends, so it is particularly important that a program of part nerships intended to support the construction of the NSDI be allowed to focus on that goal. The federal government has many other goals and objectives for its geospatial data activities besides the promotion of the NSDI. Geospatial data are used for many purposes, and their use supports many goals. As a result, there is some danger that programs designed to promote the NSDI may become convolved with other programs, be diverted to serve other needs, or expected to serve too many different purposes. At the same time, it must be recognized that many projects and programs depend on accurate and current spatial data and the cost of creating and maintaining the data is a legitimate cost item (OMB, 2000; see Box 3).

BOX 3 NSDI AND THE OFFICE OF MANAGEMENT AND BUDGET

It must be recognized that the activities of the FGDC partnership programs were not designed to be a panacea for solving all problems associated with sharing spatial data. Successful models rely on a combination of organization and financial resources. Over the past two years, the Office of Management and Budget has taken a keen interest in NSDI issues. In July of 2000 it held a GeoSpatial Information Roundtable with the objective of identifying the financial and institutional barriers that impede development of the NSDI. This meeting was attended by 110 senior representatives from various sectors. This gathering recognized the importance of the NSDI to E-Government and E-Business, and highlighted FGDC's role in its stewardship. While the OMB objectives in this sphere parallel those of the FGDC, a report Collecting Information in the Information Age (OMB, 2000) argued for a new paradigm that would build the NSDI from the "bottom up". The report recognizes the importance of scale, and notes that "State, local and tribal entities will build much of the NSDI... The challenge for the Federal government is to leverage this investment, coordinate efforts, and help state and local governments and the private sector make the data available regionally and nationally" (OMB 2000). The OMB report also recognized that "By itself, FGDC's resources are insufficient to steward the building of 'natural clusters' of partners." The participants in the roundtable developed a set of recommendations that emphasized many of the same issues that the MSC has addressed in several of its reports. For example, it also advocates the development of an extended framework.

The OMB initiative established a model for Implementation Teams (I-Teams). These teams develop comprehensive plans, conduct needs assessments, and formulate implementation strategies. Although this approach was only publicized in the summer of 2000, within a year I-Teams had been established for Arizona, Arkansas, Nebraska, Delaware, Kentucky, Wayne County Michigan, New Jersey, New Jersey and New York Metro Region, New York City, Texas, Utah, North Carolina. Montana, and Oregon. Some of these initiatives are already quite extensive. For example, the Utah Framework Implementation Plan is a comprehensive assessment of statewide needs and a blueprint for creating the NSDI within the state. It is very clear that the OMB initiative is providing a valuable umbrella for coordination. According to the Utah plan, "The OMB Information Initiative to align the needs and resources to continue to develop the National Spatial Data Infrastructure provides public and private agencies in Utah an opportunity to focus on mutually beneficial partnerships. The results of these efforts will help to provide integrated information for analysis of issues and decision-making at federal, state, local, and Tribal levels of government. Further it will provide a common frame of reference for communicating information and concepts of complex issues to citizens...."

The OMB initiative also called for the establishment of a 'financing solutions team' that would examine ways to reconcile the need for long term capital financing and the reliance on short-term annual funding mechanisms. As a consequence of this suggestion, the FGDC sponsored a report, Financing the NSDI: National Spatial Data Infrastructure—Aligning Federal and Non-Federal Investments in Spatial Data, Decision Support and Information Resources. Revision 2.0 of this report is now available for public comment.

3

Future Partnerships and the Evolution of NSDI Activities

The NSDI, and the partnership programs that have been an integral part of it, clearly need to move beyond the stage of evangelizing the concept of the NSDI, promoting its goals, and demonstrating its possibilities. Looking forward, both the NSDI and its partnership programs need to move rapidly on to new and enhanced efforts aimed at fulfilling the key objectives of the NSDI; specifically, to:

1. Populate the Framework database in a truly sustainable production mode rather than as isolated experimental or prototype project;
2. Develop and disseminate the procedures and technologies needed for effectively and efficiently building, maintaining, integrating, distributing, and using the data;
3. Continue the process of establishing clearinghouses and promulgating the necessary standards to support the NSDI.

This chapter explores the further evolution of partnerships in fostering the adoption of the NSDI. The success of future partnerships should be assessed by determining, in a rigorous fashion, how these efforts (and therefore the NSDI itself) have reduced redundancy in geospatial data collection and maintenance; reduced overall costs in performing these tasks; improved access to geospatial data; and improved the accuracy of the data used. The attainment of these tactical goals are very closely related to the Framework data efforts. These efforts support the other, less tangible but broader and strategic objectives such as increased citizen participation in decision making, and the provision of improved information to support decision making at all levels of government as well as in the private sector. The evolution of Framework-related NSDI activities and the supporting partnership program can be divided into two major categories of activities: Framework data production; and data access, use and other Framework issues.

FRAMEWORK DATA PRODUCTION

The challenge in this area is to make effective use of partnerships to stimulate, encourage, and enable the shift from small-scale, project-based data creation and maintenance efforts to large-scale, sustained, and efficient data creation, integration, and maintenance. Because Framework data are, by definition, fundamental to a broad range of geospatial information applications, it is a core goal of the NSDI to ensure that these data are being produced and maintained. Since the operating premise of the NSDI is that state, local, and tribal governments as well as private sector and NGO entities are each potential key contributors to the Framework, their successful participation in data production is a requirement for the success of the Framework and hence the NSDI itself. Therefore, because of budget constraints, partnership programs must take all possible steps to ensure that the Framework is, in fact, being populated and maintained. With the understanding that the federal government is not in a position actually to fund full-scale, ongoing Framework production efforts across the range of non-federal, data-producing organizations, how can the federal government use partnership programs to address the Framework data production goal most effectively?

- *Increase the scale, scope, and accountability of partnership activities.* This could be accomplished by selecting a small number of key non-federal entities that would be willing to participate in carefully monitored and documented data production and maintenance tasks for specific Framework layers. The objective use would be to rigorously evaluate specific approaches to data capture and data update. These experiments could be based on the use of new technology or an evaluation of protocols and procedures. The selected partners would be evaluated on their willingness to establish the capabilities to measure cost savings, data access improvements, and data accuracy increases, etc. The goal of these activities would be to take the technological and organizational steps required to put in place a complete Framework data production system, and then to run this system for sufficient time to obtain measurable and statistically significant assessment results. If the data production activity is determined to be a success, based on the criteria listed above, the goal would be to then clone the system nationwide, to the degree appropriate. Each of the partnership projects would be evaluated against the four key criteria: reduced redundancy, reduced costs, improved access, and improved accuracy. Not only should partnership programs explicitly require the capture of these factors on a before-and-after basis, but also steps need to be taken to assist non-federal organizations to take advantage of proven methods to achieve these goals. Once the assessment results indicate that these goals have been achieved, the technological and organizational or other aspects of a production system would be disseminated to the community. This process would significantly enhance the ability of non-federal organizations to produce and maintain Framework data in a manner that has been shown to be effective and efficient (e.g., soil data in Minnesota; see Box 4 below). The FGDC could also identify additional successful case studies where federal funding has resulted in partnerships that have benefited both the federal and non-federal organization. These case studies could be compiled into a "cookbook" that would provide guidance to others. A good example of such a resource is the recently completed NSDI Communications Toolkit. These communication tools were developed through a cooperative partnership between the National States Geographic Information Council (NSGIC) and the FGDC (available at: http://www.fgdc.gov/nsdi/docs/communications/index.html).

Of equal importance is the development of software tools that facilitate the integration of data from a variety of sources. In fact, the vendor community has made remarkable progress in this area. Efforts such as Microsoft's Terraserver clearly demonstrate that users can use their Web Browsers to easily access a huge quantity of image and other spatial data sources. The efforts of the Open GIS consortium's Web Mapping Test Bed demonstrate that similar tools can be used to combine data housed on several different FGDC clearinghouse nodes. Such nodes now number more than 240 located in 26 different nations. There has also been considerable technological advancement integrating desktop GIS software with industry standard database management systems and common office products. Better support for the development and use of metadata has facilitated easier exchange of spatial data among formats, map projections, and datums. In fact, map projections can be converted "on the fly" and several databases can be integrated into a single project. Software wizards and improved on-line help systems have led to significant improvements in the usability of sophisticated spatial analytical tools.

Partnership programs designed to support this kind of complete production system and evaluation effort will necessarily require a higher level of per-project funding than has been available in previous partnership programs. Clearly, the more of these efforts that can be funded, the more rapidly the successful population of the Framework database will occur.

- *Identify whether critical components of the Framework database are being adequately addressed,* either by the federal agencies or by non-federal organizations, and take action to address any gaps that are identified. Such gaps may be geographic in nature, thematic, scale-specific, etc. A strategy for addressing such gaps may include providing incentives to an organization to perform the data production, even though the organization would not normally produce such data. In the extreme, it may be determined that it is in the broad public interest to ensure that these data exist and are maintained, and therefore that subsidies or outright funding of the activity might be appropriate. Based on the specific Framework layer(s) involved, one or more federal agencies may have a particular interest in ensuring that the data are collected and maintained and therefore may support the activity financially, or alternatively may collect the data itself.
- *Offer creative incentives* for non-federal organizations to carry out their Framework data production and maintenance missions. These incentives could include cash awards based on completion and continuing maintenance of Framework data. Such incentives could be

BOX 4 SOIL DATA IN MINNESOTA-A PARTNERSHIP SUCCESS STORY

A 1994 survey of the Minnesota GIS community identified soil data at the top of the list of needs for new and improved data. The need was especially high for county governments and natural resource agencies. At the time, only one of Minnesota's 87 counties had a spatially correct digital soil map and the rate of production for such products by the Natural Resource Conservation Service (NRCS-then called the Soil Conservation Service) was one county per year. Something needed to be done to provide the required geospatial soil data.

NRCS recognized a need to accelerate soil mapping nationwide and joined with USGS and other federal agencies to create a National Digital Orthophoto Program, with the expectation that the resulting orthophotos would provide a solid base for creating new soil maps. The Minnesota legislature provided matching funds, which accelerated completion of orthophotos across the state. As a consequence of the availability of these orthophotos, NRCS scientists focused new mapping activities on Minnesota.

The Minnesota Governor's Council on Geographic Information created a soil committee, which studied the situation and determined that the biggest problem for many counties was spatial distortion in many of their soil maps, caused by lack of an orthoimagery base when the maps were compiled. Most Minnesota counties are in areas of low to moderate relief, and there was hope that these existing soil maps could be adjusted to the spatially correct orthophotos using elevation data collected as part of the National Digital Orthophoto Program.

The Minnesota Legislature, using special funds set aside for investing in natural resources, funded research by Professor Jay Bell at the University of Minnesota to see if such adjustments could be made without distorting other parts of the map. The project was successful and his approach is now being considered for approval by NRCS for use in other states. The approach has also helped focus fieldwork in counties updating obsolete soil maps.

As of late 2000, fourteen Minnesota counties have spatially-correct, modern digital soil maps and ten more are in progress. This progress would have been impossible without the contributions of the NRCS, USGS, the state policy council, the state legislature, individual counties, and the University of Minnesota.

contingent upon demonstrating achievement of the four assessment criteria and continued improvement over time. Other incentives could include access to NSDI software tools, applications software, training materials, etc.

DATA ACCESS, USE, AND OTHER FRAMEWORK ISSUES

In addition to the core goal of populating the Framework, several other critical issues must be addressed in order for the NSDI to be a success, especially in regard to the broader objectives of improving decision making through increasing the effective use of geospatial data at all levels of government, by citizens, and in the private sector. As with Framework data production, partnership programs are needed to address these issues effectively, since both the definition of potential solutions and the implementation of the solutions need to occur in the geospatial community at large.

• *Data integration* (vertical and horizontal). If the enormous potential benefits of the NSDI are to be realized, datasets produced by different organizations, covering different themes and geographic areas, and at different scales, must be used in conjunction with each other, as well as with non-Framework datasets. While the focus on transfer standards and metadata standards has been a necessary step in realizing true integration, it is now necessary to use the standards to actually achieve integration. Vertical integration ensures that data elements from disparate themes over the same geographic area demonstrate logical and geometric consistency. This is difficult enough to achieve when integrating data of the same structural type (e.g., vector hydrographic data and vector transportation data), but becomes even more problematic when integrating disparate types (e.g., raster orthoimagery, matrix-based terrain elevation data, and vector data). Nevertheless, even though these themes may be produced by different organizations, they must be technically compatible in order for the full benefits to be realized. A critical component of this type of integration will be the acceptance of standards that are being developed by the FGDC subcommittees. For example, the Ground Transportation Subcommittee's proposed standard on transportation features provides a model for different organizations to refer to the same road segment and assign attributes that meet their specific needs.

Vertical integration issues can be addressed through several approaches. First, at the time of collection or production, steps can be taken to help maximize consistency across themes. Given the "ground truth" of orthophotography (assuming accurate positional control, sufficient resolution, etc.), production systems that incorporate this imagery into the process (e.g., as a backdrop if not for actual extraction of features) may ensure a level of consistency and accuracy across themes. Similarly, the collection of elevation models utilizing hydrographic information not only improves the accuracy of the terrain data but also helps ensure consistency between these two layers. Thus, data production methodologies can help mitigate the vertical integration problem.

Once data have been collected, tools, both automated and interactive, can be used for after-the-fact data accuracy and consistency checking and clean-up. Again, the development and application of these tools can improve integration. Finally, at the user end of the spectrum, that is, in applications, analytical procedures, visualization tools, etc., smart software can deal with potential integration problems and still allow for appropriate use of the various data themes together. For example, inconsistencies or other problems may dictate that it would not be appropriate to combine two themes in an analytical overlay fashion (e.g., point-in-polygon or polygon overlay calculation), even though visual integration at a particular scale is perfectly legitimate. The Web Mapping Testbed of the Open GIS Consortium (OGC, 2001) addresses this issue directly, and has already demonstrated substantial success.

Horizontal integration refers to the simultaneous use of datasets across collection or jurisdictional boundaries. This is key, for example, in dealing with issues on a regional (e.g., river basin or watershed) basis. The logical subdivisions fall into three administrative levels: regional, state, and national. The level of responsibility and authority that resides at the state level varies considerably from state to state. For example, the Texas Natural Resources Information System provides public domain statewide coverage of a number of the framework datasets at a scale of 1:24,000 and quarter quad digital orthophotographs. Other states, such as

Tennessee, have made a major commitment to providing high resolution (1:48,000 and 1:1200) framework data and tax parcels on a statewide basis. The Tennessee state government is providing 75 percent of the $30 million for the project (ESRI, 2001b). Clearly, such a state level commitment will greatly facilitate horizontal integration. As with vertical integration, the problem can be addressed at each stage of the geospatial data collection, production, and use spectrum. To the extent that national-level base data, even at smaller scales than optimal, can be used to identify tie points at the boundaries between data coverage areas, some inconsistency can be avoided. Following data collection, software for checking the consistency of data across collection boundaries can be used to detect inconsistencies and either resolve them in some automated fashion or flag them for manual clean-up. Finally, at the data use or application stage, appropriate use of data across collection boundaries can be accomplished, even in the presence of anomalies such as gaps, attribute changes, and other inconsistencies.

Thus, to deal with the integration problem, the NSDI and future partnerships should address the issue by encouraging integrative actions at each stage in the geospatial data process. This may include the development of procedures, processes, software tools, standards, guides, and other aids.

• *Data use and applications.* Clearly, the true payoff of the NSDI will be closely tied to those geospatial data-based applications that make use of Framework and other data to address specific problems or issues facing governments, companies, and NGOs. In the next stage of the NSDI partnership program, the development and diffusion of geospatial applications will be critical to the perceived success of the entire effort. Applications of geospatial data in a particular domain involve the integration of domain-specific data with Framework data, mapping, and visualization of the data, geometric processing, spatial search and retrieval, tabulations of summary data, and incorporation of the data into analytical or predictive models. The ultimate objective of the utilization of geospatial data and technologies is to promote and enhance information-based decision making. There is widespread recognition that spatial data are the core of a new level of services to the citizens. Taxpayers have similar expectations from their local governments as they do for other information-based services provided by their banks, travel agencies, bookstores etc. These E-government solutions require considerable innovation to bring a high level of web-based services to a relatively unsophisticated user community.

The University Consortium for Geographic Information Sciences (UCGIS) presented a good overview of critical application domains (UCGIS, 2000b). The domain areas include:

1. crime analysis,
2. emergency preparedness and response,
3. transportation planning and monitoring,
4. public health and human services,
5. urban and regional planning,
6. water resources, and
7. involving the public.

This list could easily be expanded to include important issues such as environmental protection, equitable tax assessment, school zoning, bus routing, hazards and risk assessment, and growth management. One of the great benefits of sharing spatial data occurs when multiple uses are realized that extend beyond the original need. The Census TIGER (Topologically Integrated Geographic Encoding and Referencing system) line files are the primary example of this. The MSC has argued that by "throwing the goodies over the fence," the Census Bureau essentially created new markets and expanded the application domains for GIS practitioners. An important consideration is whether greater cooperation could result in better data with additional attributes or improved spatial resolution for essentially the same cost.

The NSDI and associated partnerships should develop application guides based on successful projects that have used geospatial data and tools to address these issues. Ultimately, these guides would be complete "how-to" cookbooks that would identify the Framework data, non-Framework and domain-specific data, and the application software tools that are available, where to get them, and how to use them. To the extent that commercial or public-domain software or data exist to address these domains, the guides would point to those resources. In the case where the needed resources do not exist, a partnership program may develop new software or, preferably, modify or customize existing software.

Assuming that these application guides would be available from the clearinghouses through the web (virtual town halls), and could include the needed tools or information on how to get them as well as the ancillary domain-specific datasets needed for application, there should be rapid diffusion of geospatial data use. Partnership programs can be designed to implement this scheme, for example, through the funding of application-specific Community-Federal Information Partnerships grants. These grants could stimulate the creation of application guides as well as the development of needed tools, software customizations, and domain-specific datasets.

THE TIME DIMENSION: DATA UPDATE, ARCHIVING, AND CHANGE DETECTION

For the NSDI to fulfil its mission, the currency of the data (particularly those themes of the Framework database that exhibit significant change yearly or more frequently, e.g., transportation and orthophotos) must be addressed to assure users that information is accurate as of a specified date, and that the information is not so outdated as to make its use in specific applications problematic. Contributors of Framework data, particularly the changeable themes, should be encouraged and assisted to maintain the data in a structured ongoing process so that some degree of predictability and confidence in terms of the utility and timeliness of key data elements will be assured. Future partnership programs should therefore provide incentives to organizations to establish systems for regular updates and maintenance of Framework data based on transactions. These transactions should include changes to the built environment, such as new roads and subdivisions. Such continuous update activities would provide critical information for emergency 911 systems, and would also significantly reduce the start-up cost for the decennial census. Update guidelines, by theme and perhaps by scale, should be developed to encourage partner organizations to adopt and commit to a regular schedule of updates, with specific definitions of a minimally acceptable update for any particular theme. For example, for the transportation theme, certain features and attributes viewed as essential to a large majority of applications would be required to be updated on an annual (or quarterly, etc.) basis, whereas other features and attributes could be updated less frequently. It should be noted that the proposed standard for transportation features accomplishes this by uniquely labeling the points and segments that comprise the road network. This enables one to unambiguously refer to a specific feature and locate it on the earth's surface. This type of national registry of transportation features would allow for continuous updates and multiple representations. A concurrent effort by the FGDC cultural and demographic subcommittee has developed a proposed standard for assigning addresses to these road segments.

It is important to note that in the 1993 report, *Toward a Coordinated Spatial Data Infrastructure for the Nation* (NRC, 1993), the MSC called for the development and maintenance of a national Street Centerline Spatial Database (SCSD). The SCSD was considered to be one of the cornerstones of the National Spatial Data Infrastructure (NSDI). In 2001, there is no clear program for a coordinated effort to maintain a SCSD. It is also not clear whether input for the continuous monitoring of its proprietary street centerline data should fall in the domain of the USGS, the Bureau of the Census, state highway departments, local governments, or private companies.

Partnership programs should be established to develop and test technical and organizational systems for data update. These should contribute to the development of standard protocols and guidelines that would facilitate a high degree of uniformity nationwide. Since many public and private organizations would benefit from accurate and timely spatial data, it is important to consider innovative approaches to public-private funding, and to devise appropriate federal financial incentives that could facilitate the continuous maintenance of framework data.

In addition to data update, earlier versions of data must be retained for change analysis and historical review, both of which the availability of datasets representing two or more time periods enables. For example, the Urban Dynamics Research Program (UDRP, 2001), a partnership of the USGS, NASA, and several universities, uses historical land-use data to model and predict urban growth in U.S. metropolitan regions. Other policy issues depend on a consistent method for monitoring changes in land use and land cover. The detection of key changes in geospatial data, the description and measurement of change, and the analysis and modeling of change are required for many applications. For example, in order to meet programmatic goals—such as: "Analyze land use change in large metropolitan areas using USGS-derived temporal data" (UDRP, 2001)—it is essential to have the supporting spatial information. Partnership programs that utilize the NSDI Framework data for change detection or analysis for specific applications would be valuable as a means to ensure that the NSDI resources support this functionality adequately, and also to develop tools and methodologies for change detection and analysis in the NSDI context. This would make an important contribution toward the goal of increasing the use of geospatial data to address real-world decision-making needs. A data archiving function could become a standard task for the NSDI organization itself (i.e., centralized archiving) or it could fall within the responsibility of the contributing partner organization. From a future partnership perspective, this issue should be addressed in order to develop guidelines and assistance to partner agencies to help ensure that data are being appropriately archived. Again, since consistency is so important, guidelines by theme (and scale) will be required, as will technological tools that can assist data contributors in maintaining old versions of their NSDI-relevant data and making them available to users. All of these extended services rest on the same basic assertions: that through partnerships, the NSDI will reduce costs and duplication and improve accuracy and access.

PRIVACY, THE PRIVATE SECTOR, AND PUBLIC ACCESS ISSUES

Because the NSDI network of information assets, including Framework data, will not be comprised entirely of public-domain data, many of these information assets will be subject to concerns about privacy, private and public rights, and free use versus pay-for-use. To exclude these latter assets would result in a watered-down, more expensive (from a taxpayer's if not a user's perspective), less detailed, less accurate, and hence less useful NSDI. Therefore, these issues must be addressed, and future partnership programs should be responsible for establishing the policies, tools, and systems needed for an NSDI that can adequately handle private-sector as well as public-sector data, confidential as well as non-confidential data, and free-use as well as restricted-use data. Partnership programs dealing with public-sector data which may have some associated privacy or confidentiality concerns (e.g., dealing with individual-level property information in a cadastral dataset) should identify or develop guidelines and tools for dealing with this issue in a way that prevents the unauthorized or inappropriate use of such data, yet still makes available as much of the information as possible without violating privacy or confidentiality guidelines.

It must be noted that many of the challenges that face the development of the NSDI based on public private partnerships are being addressed by the private sector. One example is MicroSoft's establishment of the Terraserver, which provided free access to federal data and images. Another example is Environmental Systems Research Institute's Geography Network. The Geography Network is "a concept that promotes the sharing and distribution of geospatial information via the Web, allowing consumers to have access to information that will allow them to understand their geography and apply this to their everyday and business use" (ESRI, 2001a). Using the fundamental power of the Internet, contributors publish links to web servers that house their geographic data. In this manner the contributor remains the custodian of the data and is free to establish its own data maintenance program. In many ways the Geography Network represents an alternative to the FGDC data clearinghouses. However, according to ESRI (2001a):

> "The Geography Network complements and supports the FGDC's efforts to create a National Spatial Data Infrastructure,...assists in building relationships among the organizations that are supporting the NSDI...[I]t provides the infrastructure to build and support the sharing of data across different industries, organizations and nations."

While still in its infancy, the Geography Network appears to have gained favor with a wide range of public and private data pro viders and represents an intriguing business model. The Texas Natural Resources Information System, the New Jersey Office of GIS, the Pennsylvania Spatial Data Access, and the USGS, for example, are all using the Geography Network to disseminate public domain data, while private data providers, such as Geographic Data Technology, are using the same system to collect fees for their proprietary data. In this manner the Geography Network is a combination of a geospatial library and an e-business venture. A unique feature of the system is that it allows for a preview of data before any fees are charged for the actual data transfer. It is interesting to note, that in the open environment of the Internet, the FGDC data clearinghouse nodes could become part of The Geography Network. Clearly, The Geography Network will enhance the concept of the NSDI by potentially providing a highly popular and robust starting point for the search of the most complete inventory of spatial data for any part of the world. The ultimate success will be judged by the public in terms of performance, completeness, and ease of use, however, it must also be noted that The Geography Network relies on contributors who have adopted FGDC metadata and content standards. It could be argued that this is exactly the type of public private partnership that will make the NSDI a reality.

Private-sector participation in the NSDI will require that guidelines be established and mechanisms put in place to manage users' access to licensed datasets or elements. Much work needs to be done first to develop the policy for integrating private-sector data within the NSDI (e.g., Will Framework data be completely public domain or free access? Can a free-access version of Framework be made available for those unable or unwilling to pay for data access, as well as a fee-based version for those who will pay? Is it possible to establish consistency in pricing policies?), and then to establish the infrastructure needed to manage access, licensing, and fees. This process should take full advantage of the ever-improving state-of-the-art in e-commerce tools, particularly those dealing with selling digital goods, such as Qpass, which manage the sale and distribution of information from other sites, such as image repositories (e.g., Corbis, 2001), financial databases, and news databases (Qpass, 2001).

To address the public-access issue, it may be feasible to implement a public-use category that is free in all cases, even when private sector data are involved (this may be particularly appropriate for Framework data). In addition, there may be added data features or attributes supplied via the private sector (or NGOs or even local governments) that require payment but are fully integrated with the NSDI. This information source, rights, licensing, and payment architecture must be defined and then implemented. Partnership programs with private- and public-sector organizations can help greatly in moving towards this goal by using real-world examples of data sources whose introduction to the NSDI will require these issues to be addressed. Policy guidelines, technology solutions, and organizational structures should each be addressed in partnership projects dealing with this issue.

THE GEODATA ALLIANCE—AN INNOVATIVE ORGANIZATIONAL APPROACH FOR DEVELOPMENT OF THE NSDI

The committee is encouraged by the efforts of the FGDC to seek creative ways for expanding the participation in the NSDI initiative and to develop a sustainable organizational structure that can build on federal efforts. In addition to the OMB Initiative to establish regional I-Teams to develop the NSDI, the FGDC has also been advocating the establishment of the GeoData Alliance. The GeoData Alliance stems from a presentation at the 1999 GeoData Forum by Dee W.Hock, founder and CEO of Visa International, and Coordinating Director of the Chaordic Alliance. The development of Visa International was based on the need to develop a functional operating organization amongst an extensive set of loosely linked activities in the marketplace. Similarly, a GeoData Alliance could create a more structured organization within what is presently a fairly chaotic and disorganized set of players in the geospatial data arena. The FGDC played a lead role is creating and supporting a broad-based drafting team to develop an organizational design for the alliance. In September 2000, the drafting team generated a report that lays out a detailed blue print for a nonprofit organization with the stated purpose to foster "…trusted inclusive processes to enable the creation, effective and equitable flow, and beneficial use of geo graphic information..." (GeoData Alliance, 2000). The overriding concept of the GeoData Alliance is "...to create contexts in which diverse individuals and institutions can come together to pursue common interests, collaborating when appropriate and competing vigorously in other ways...." (GeoData Alliance, 2000). An integral aim of the Alliance is to collaboratively develop strategies and plans for the realization of the NSDI. It is interesting to note that one of the recommended practices is to create or support transactional systems that would focus on framework data. At the same time, one of the guiding principles in the creation of the GeoData Alliance is that "Geographic information has inherent value and the creators of that value should be equitably compensated." This raises perhaps the most difficult obstacle that the NSDI concept faces. Increasingly, equitable compensation takes the form of a licensing agreement for the restricted use of the data from a private vendor or a license from a public agency that is trying to recoup its capital investment. Such licensing agreements are generally in conflict with policies of the U.S. government agencies that traditionally have acquired outright ownership of data. By acquiring ownership, government is able to offer broad access to citizens and the commercial sector to the data it acquires, as well as access to any derived public records. If government licenses rather than purchases data from the private sector, many of these benefits are threatened. Once government begins to acquire information resources by license, it will be forced to license out or contractually limit dissemination of the data. Therefore, the principle of equitable compensation to data creators becomes thorny.

The GeoData Alliance is open to individuals, organizations, and other alliances, and is governed by a council of 32 trustees. In the committee's view, the creation of the GeoData Alliance is a significant step in the evolution of the NSDI and the role of the federal government. It is not clear how the concept of the GeoData Alliance will be reconciled with the OMB-supported I-team initiatives that have already proven to be extremely popular. Although the FGDC has played a significant role in fostering the development of the GeoData Alliance, the committee foresees that the creation of such a nonprofit organization could surrender the preeminent role that the FGDC has played in NSDI activities to date. Since the concept is still in its infancy, it is not clear how the different sectors would interact within such an organization. For example, considerable attention should be paid to the balance of power. If it is dominated by the private sector, such an alliance could disrupt the sharing of data that has been a cornerstone of the NSDI concept.

It is also important to note that the FGDC has been playing a major role in promoting global data sharing. It has participated in all five Global Spatial Data Infrastructure (GSDI) conferences, and serves as the organization's permanent secretariat. Although the GSDI is still a fledging concept, is significant that 43 countries recently sent representatives to Cartagena, Columbia, to the Fifth Global Spatial Data Infrastructure Conference. The FGDC staff provided considerable assistance in the development of a "cookbook," *The Spatial Data Infrastructure Implementation Guide,* and support for the GSDI website (GSDI.org). This cookbook provides extensive guidance and recommendations regarding policies, organizational principles, and standards. Indications are that the FGDC involvement in the GSDI setting will lead to a more coherent organization of several of the nation's international spatial data efforts such as Digital Earth, Global Map, the Global Disaster Information Network, and the United Nations Environmental Programme.

4

An Extended National Spatial Data Infrastructure Framework: The Role of Other Organizations

In 1993, 1994, and 1995 this committee issued reports on the contents of NSDI, the need for development of a robust NSDI in the United States, and a method for satisfying that need through creative partnerships (NRC, 1993, 1994, 1995). These reports have received widespread acceptance, and as the concepts embodied in these reports mature, it is becoming increasingly obvious that an effective and widely used NSDI will be developed with substantial if not primary input from organizations outside of the federal government. The FGDC has undertaken the task of promoting the development of the NSDI. The core contents of the NSDI are referred to as the Framework. The seven themes that form the Framework for the NSDI were detailed in Chapter 1 and are identified in Table 1. In addition, the Framework also includes procedures, guidelines, and technology to enable participants to build, integrate, maintain, distribute, and use Framework data. In this chapter we explore the roles of non-federal organizations, and offer suggestions as to appropriate extensions of the NSDI conceptual framework at the state, tribal, city, and county levels.

ARGUMENTS FOR AN EXTENDED FRAMEWORK

The Framework, as it is now defined, serves the purposes of the federal government and is useful for national or large-region proj ects. But the seven themes, as now defined, may not fulfill the needs for more local studies by states, tribal nations, cities, and counties, for two reasons:

1. The information the seven themes encompass is required in greater detail at the local level. For example, roads may have to be described by their edges instead of by their centerline. Property owners and local officials often need to define and locate the right of way between an individual's property and a road. Tax maps are often the most critical resource in resolving local land use and zoning conflicts. These maps must also be integrated with the location of specific buildings and the location of utility infrastructure networks. It simply is not feasible to accurately depict these features at the map scale used by federal mapping organizations. In fact, the largest scale federal map series is still the USGS 1:24,000 series of 7.5-minute topographic quadrangles. In the local mapping community, maps of this scale would be considered small scale with a map accuracy of approximately 40 feet, based on the statistical methods advocated in the FGDC standard for specification of spatial accuracy. The base maps for large-scale mapping are often legally required to be of a scale of 1 inch to 100 feet or 1:1200.

   The requirement for large-scale source materials is critical for the development of federal-local partnerships. It must be noted that this is not the first time that a NRC committee has highlighted the need for federal support for the development of a nation-wide database that accurately depicts individual property ownership records (see Box 5). The committee is pleased to note that the FGDC has recognized this need for increased resolution, concluding in its 2000 assessment of the Community Demonstration Projects that "...many federal datasets lack sufficient resolution to support local planning needs..." and advocating that "...federal agencies should continue to enhance the quality of data using the latest technology..." (FGDC, 2001).

2. Additional themes may be needed at the state, tribal nation, county, and city levels: for example, water rights in the western United States, or utility information at municipal levels. It is clear that not

all types of data layers are used by everybody, but at the state, tribal nation, city, and county levels some additional themes are used by a great number of users. In such cases, it may make sense to incorporate these additional data themes into an extended Framework, incorporating all fundamental data layers identified for the cities, counties, tribal nations, states, and the nation.

In general, one would expect that data layers might require increasingly finer resolution and perhaps a greater amount of data detail at the city or county level than at the state or tribal nation level. The same may be true of the state or tribal nation level compared to the national level. Of course, some data layers may have the identical resolution and data detail in more than one of the three geographic levels (nation, state or tribal, local). The committee developed a matrix that attempts to examine the responsibility for the creation and maintenance of different framework data layers (Table 1). The data layers are the ones mentioned in the National Academy of Public Administration's 1998 publication, *Geographic Information for the 21st Century* (NAPA, 1998). The intent of this matrix is simply to demonstrate that the NSDI must be built on the basis of shared responsibilities, costs, benefits, and control. The committee recognizes that responsibilities will vary across the country depending on available resources and differing mandates and regulations, as well as property ownership, density of development, and other factors.

The matrix could serve as a useful starting point for the development of an extended framework. Preliminary designations of primary and supplementary responsibilities for each layer are indicated. It should be noted that orthoimagery is viewed as a critical component of the development of any extended Framework data collection effort. The ultimate responsibility for the creation and maintenance of any individual theme would be determined by the legislative or regulatory mandates in a particular region. It must be acknowledged that local government requirements for zoning, property assessment, or other land-use decisions will often determine where such authority resides. Some local governments have been able to couple these mandates with the requisite financial resources to develop such systems independent of other organizations. The chal

> **BOX 5 1980 NRC REPORT-SOME FINDINGS AND CONCLUSIONS REMAIN RELEVANT TO NSDI PARTNERSHIP PROGRAMS IN 2001**
>
> In 1980, the NRC Committee on Geodesy commissioned the Panel on a Multipurpose Cadastre to produce a report entitled *Need for a Multipurpose Cadastre* (NRC, 1980). Twenty years later it is useful to revisit some of the findings and recommendations contained in that report:
>
> "There is a critical need for a better land-information system in the United States to improve land-conveyance procedures, furnish a basis for equitable taxation, and provide much-needed information for resource management and environmental planning."
>
> "The major obstacles in the development of a multipurpose cadastre are the organizational and institutional requirements. Reorganization and improved quality control for existing governmental functions will be required. Each of the components of the cadastral system already exists somewhere within our existing governmental structure. Many of the required data are being generated at the local level, and in most cases the users are the individual citizens and the local government officials and planning organizations."
>
> "The components of a multipurpose cadastre are the following:
>
> 1. A reference frame consisting of a geodetic network;
> 2. A series of current, accurate large-scale maps;
> 3. A cadstral overlay delineating all cadastral parcels;
> 4. A unique identifying number assigned to each parcel that is used as a common index of all land records in information systems; and
> 5. A series of land data files, each including a parcel identifier for purposes of information retrieval and linking with information in other data files."

AN EXTENDED NATIONAL SPATIAL DATA INFRASTRUCTURE FRAMEWORK: THE ROLE OF OTHER ORGANIZATIONS

> The Panel recommended:
> "...that technical studies continue to be sponsored by the federal government to identify consistent land information and display standards for use among and within federal agencies and between federal and state governments. These studies should rely on the authority of state governments to adopt the standards and organize the data collection, in cooperation with the federal government to ensure compatibility on a national basis, delegating these functions to local governments where appropriate.
> ...that each state authorize an Office of Land Information Systems, through legislation where necessary, to implement the multipurpose cadastre.
> ...that local governments be the primary access point for local land information."
> "We recommend support by the federal government for the establishment of a center or centers of excellence in land-information science, for the purposes of providing a program that develops scholars and professionals. The curriculum should include direct experience with land-data-systems problems."
> The present committee notes that although there has been some organizational progress since 1980 (e.g., NSGIC, FGDC, GeoData Alliance), the fundamental need to improve the nation's geospatial data capabilities and resources remains as a challenge to the implementation of a robust NSDI.

lenge is to find ways to reach a common ground that can benefit all the potential users. This visual representation of the actual features on the ground, in their planimetrically correct position, provides the best evidence and source material for updating and correcting spatial data. A fundamental goal and driving force behind an extended Framework is that data will be collected once and maintained regularly. In other words, if a data layer is part of the NSDI and also a component of both

TABLE 1 Potential Responsibilities for Data Layers in a Spatial Information Infrastructure

Theme	Federal	State	Local
geodetic control	**primary**	**supplementary**	**supplementary**
cadastral data	**supplementary**	**supplementary**	**primary**
political boundaries	**primary for states and international**	**primary for counties and state reserves**	**primary for municipalities and local areas**
base cartographic and elevations	**primary for scales smaller than 1:24,000**	**supplementary for road building and state projects**	**supplementary for local projects**
bathymetric	primary for offshore areas, int'l waters	supplementary for lakes and reservoirs	supplementary for ponds
geologic	primary	supplementary	supplementary
hydrography	**primary**	**supplementary (water rights)**	**supplementary**
transportation & utilities	**supplementary**	**primary for highways**	**primary for some utilities**
soils	primary for coordination	supplementary	primary for survey
vegetation	primary for federal lands	primary for state lands	primary for local lands
wetlands and wildlife habitat	primary	supplementary	supplementary
cultural and demographic	primary	supplementary	supplementary
digital orthoimagery (scale dependent)	**primary at coarse resolutions**	**supplementary**	**primary at fine resolutions**
statistical base maps & address files	supplementary	supplementary	primary
land cover & land use (added to napa list)	primary for land cover	supplementary for both	primary for land use

NOTE: **Bold** text boxes include the seven NSDI Framework themes.

AN EXTENDED NATIONAL SPATIAL DATA INFRASTRUCTURE FRAMEWORK: THE ROLE OF OTHER ORGANIZATIONS

a State Spatial Data Infrastructure (SSDI) and a Local Spatial Data Infrastructure (LSDI), the data for these layers need to be collected at the lowest level and generalized to the other levels. This ensures logical consistency among the parts of the extended NSDI Framework. Again, it must be noted that the data content standards being developed by the FGDC working groups are facilitating this process. The 16 accepted standards and the additional ones under development represent a major effort to develop consistent definitions and descriptions of geographic features and attributes.

There are at least nine major steps necessary to realize this extended Framework:

1. Definition of the contents of the city, county, or local extended Framework.
2. Definition of the contents of the state or tribal nation extended Framework.
3. Definition of the extended Framework hardware architecture.
4. Definition of coordination mechanisms.
5. Assignments for layer responsibilities.
6. Definition of quality standards (collection and maintenance) and procedures for the development of the extended Framework at all levels.
7. Data generation in agreement with the corresponding Framework.
8. Data maintenance program.
9. Budget allocation.

This chapter primarily addresses the first and second items above. For further details and discussions, the interested reader is directed to the recent National Academy of Public Administration's volume entitled *Geographic Information for the 21st Century* (NAPA, 1988). With respect to the last item above, the lack of financial resources will be an impediment to development of an extended Framework for smaller counties, cities, and possibly states. In such cases, substantial subsidies will be needed from higher levels of government, unless development can be financed through partnerships with other organizations. .

DEFINITION OF A CITY OR COUNTY EXTENDED FRAMEWORK

The starting point for any city or county extended Framework is FGDC's Framework. Therefore, a county Framework should include geodetic control, orthophoto imagery, elevation, transportation, hydrography, governmental units, and cadastral information. The geodetic control may be supplemented at the local level by local surveys, and the orthophoto imagery could also be supplemented by larger scale coverage than that collected federally. Information on the utility location is important at the local level, and is likely to become more important as the utility industry and public-sector utility services exploit new technologies that require more accurate geospatial data. More detailed elevation data may also be part of the local jurisdiction's contribution to an LSDI. For example, we already have counties that have 0.5-foot contours derived from orthoimagery produced by the private sector under contract. For the transportation layer at the county level, it is expected that transportation features such as roads will be defined by their edges, and maybe by the spaces corresponding to the road right-of-way in addition to the road centerlines. For hydrography, additional information such as the location of each bank of the watercourse, its navigability for small craft, intakes from rivers and streams, and inputs into the same, may be monitored. For these federal Framework themes it is clear that local level data will enrich most of the layers of the NSDI.

A major difference between the local Framework and national Framework is the definition of the content for both the governmental units layer, which accurately depicts a wide range of administrative unit boundaries, and the cadastral information layer, which depicts the legal boundaries of parcels of property ownership. Whereas the cadastral information overlay from the federal NSDI could be expected to include both the Public Lands Survey System (PLSS) used in the western states and federally-owned lands managed by the Bureau of Land Management, the Park Service, and other federal agencies, the local Framework would include details of privately-owned parcels. This is an entirely different magnitude of data collection compared with the supplemental information described in the preceding paragraph. Similarly, whereas the NSDI contains international and state boundaries, the preponderance of other boundary information would be provided to the federal level by state and local levels. Municipal boundaries, voting districts, city wards, county or municipal parks, school attendance areas, and similar administrative boundaries should be provided by the local level. In some cases, the responsibility to collect, integrate, and maintain the data theme lies with the state but has been delegated to the local level. Examples abound of existing files that contain such digital information. Geospatial information describing ownership boundaries and structure footprints is often accompanied by owner name, street address, assessed valuation, and many more attributes (some databases include and make available as public domain data, square footage of buildings, floor plans, number of bathrooms, etc.).

A second difference is the reference system, although hopefully this difference will be temporal in nature. Even though it is highly recommended that NAD 83 (North American Datum, 1983), NAVD 88 (North American Vertical Datum, 1988), and latitude and longitude be used as the basis of a positional system in the NSDI, at the local government level this may not be practical. For example, most local surveys are conducted in the State Plane Coordinate system (SPC). Therefore, it may be preferable to use SPC rather than latitude and longitude for some implementations at the local level. The fact that transformation equations exist between the different SPC zones and latitude and longitude lessens the practical impact of this difference. It may eventually mean that the data are available in latitude and longitude but that a separate file in SPC is kept for local daily use. Recent developments in GIS technology allow differences in projection and datum to be overcome "on the fly."

Another major change at the state and local levels would be the inclusion of additional themes. For example, an additional theme at the local level could be the location of public services: schools, hospitals, police and fire stations, etc. Each of these features may be annotated with attributes at a level that could not be done by a national agency, yet the information would be valuable at any level. It is assumed here that the positional and attribute resolution of the data layers at the county level will be the highest (or at least no less than the resolutions at the state or national level). Traditionally, soils data have been collected at the county level. Traffic accidents and crime statistics are collected locally. Incidences of disease data are most useful at the local level. Other possible themes include ZIP code areas, zoning requirements, and traffic flows.

It is evident that a local extended Framework must be defined with the cooperation of city and county officials, and that only those additional themes used for the majority of applications should be incorporated into an extended Framework. To do this, county officials need to be involved in the discussions leading to the definition and establishment of an extended Framework. These discussions should take the form of a nation wide needs assessment which would develop a clear articulation of the content and necessary scale of spatial data required to meet specific objectives and mandates at each level of government. The outcome of this must be a list of themes and their content that can be applied at the local level. This bottom-up approach is in line with the I-Team initiatives advocated by OMB. The committee is encouraged that the National Association of Counties (NACo) began formal cooperation with the FGDC in 1997. This cooperation needs to be continued with specific goals established relating to the definition of an extended Framework.

DEFINITION OF A STATE OR TRIBAL NATION EXTENDED FRAMEWORK

The starting point of a state or tribal nation extended Framework is also the FGDC's Framework. Therefore, a state Framework will include geodetic control, orthophoto imagery, elevation, transportation, hydrography, governmental units, and cadastral information. The geodetic control, elevation, and orthophoto imagery layers may be supplemented by the state. Governmental units, a state responsibility that is often delegated to the local level (municipal boundaries, school district boundaries), would probably not receive much additional supplementation except for such features as state legislative district boundaries, state parks, and state forests. Similarly, the cadastral layer augmentation at the state level might be limited to state-owned lands, however, some states such as Maryland maintain tax parcels on a statewide basis.

A tribal nation Framework would differ from state Frameworks in several ways. Among the most important is the complex pattern of property ownership on many reservations, with some property held by the community, some by individuals, and some by non-tribal owners. This makes distribution of income from tribal assets (e.g., oil and gas lease income) particularly difficult. The involvement of the federal Bureau of Indian Affairs adds an additional bureaucratic layer that makes geospatial data management somewhat more difficult.

One major difference between a state and tribal nation Framework and the FGDC Framework is the definition of the content of the transportation layer. For example, at the state level linear transportation features such as roads may still be defined by their centerlines (as in the federal contribution to the NSDI), but they may carry additional information (county limits, mileage, snow removal, signage placement, and other maintenance responsibilities). New technologies, such as GPS-equipped vans, roadway sensors, high-resolution (1-meter) remote sensing, and digital photogrammetry, are revolutionizing the availability of accurate geospatial data in the transportation layer. In most states, departments of transportation are major agencies that handle such services as driver licensing, vehicle title and registration, interstate commerce taxes, in addition to the features listed above. As the spatial dimensions of these layers become increasingly in demand, the states will find that this information should be made compatible with the SSDI.

Hydrography is also of major concern at the state level, and includes navigation, energy, and recreational users as well as point and non-point pollution sources. There are also regional concerns over water rights and quality. Watersheds often contain several local jurisdictions, and therefore the state must assume responsibility for data relating to drainage basins. The state that handles fishing licenses typically designates public access points to lakes and waterways, patrols open water, and plays a major role in the mitigation of natural disasters involving its watercourses.

Even though it is highly recommended that NAD 83 and NAVD 88 and latitude and longitude be used as the bases of a positional system at the county or local level, this may not be practical at the state government level. For example, in the case of the State of Ohio, NAD 27 is the basic reference system for horizontal data for a large amount of the existing spatial data for the state. Some states mandate the use of the SPC. Another major difference is theme related: The location of wetlands, ecosystems, land cover, watersheds, and geologic formations are themes a large number of state agencies use. In some cases historic buildings, monuments, or burial grounds are state themes as well as features required for disaster preparedness and emergency response.

It is evident that a state or tribal extended Framework must be defined with the cooperation of state or tribal officials, and that only those additional themes used for most of the state or tribal agencies should be incorporated into an extended Framework. A meeting of the major stakeholders concerned with geographic data layers at the state and tribal level needs to be convened in order to discuss and define an extended Framework. As in the case of counties, the outcome of this step must be a list of themes and their content.

The FGDC understands that it must develop effective coalitions with state and local government organizations if it is going to succeed in the development of an extended Framework. The committee is especially encouraged by the efforts to establish a strong working relationship with the National States Geographic Information Council (NSGIC), and considers that this is the primary partnership needed to undertake the definition of the extended SSDI.

SUMMARY OF SPATIAL DATA THEMES

Many data themes have been mentioned in the above short discussions. The responsibility of the different levels of government for the various themes are described in Table 1. Where the federal government bears primary responsibility, the supplemental collection at the state and local government level must at least meet the federal data standards. In most instances, state and local standards are more rigorous than those at the federal level. But where the primary responsibility resides with the local government, supplemental information collected by state and federal governments should at least meet the local standards.

In practice this has not always been the case. For some layers, primary responsibility is shared among two or three levels of government for different parts of the layer (e.g., political boundaries and vegetation).

Data standards are a critical element of this effort. It has also become clear that accurate, current orthrophotography is a critical building block. Clearly, the federal government has a primary responsibility for a digital imagery data layer that covers the entire country. Hydrography, wetlands, and wildlife habitat, vegetation, geology, and bathymetry for offshore areas may be partially collected using imagery and the federal government has a primary responsibility in each of these areas. State and local governments have primary interests in transportation and utilities, soils, vegetation, and for certain features that can partially be collected with the aid of imagery. Cowen and Jensen (1998) have documented the spectral, spatial, and temporal resolution requirements for different types of features. However, local and state governments have different responsibilities for data layers that cannot be collected through the use of imagery. These differences between national, state and local needs are in some cases fundamental.

At the start of a new century, most jurisdictions find a plethora of data available. Of greater need are personnel and processes to assess those data and to define the form in which those data are needed at each level. Free data are not free if the user must invest thousands of dollars to use them. Until the personnel at each user interface are hired and dedicated to identifying both data needs and the processes to create the forms in which those data can be easily used, one cannot answer the question, What data are needed? We have developed a tremendous capability to collect data, driven primarily by the development of technology that can automatically collect them. We need now to develop the comparable capabilities to process, assess, and use those data. The Framework concept and its extension at the state, tribal nation, city and county levels outlined above begins this process.

ROLES OF PRIVATE INDUSTRY AND NON-PROFIT ORGANIZATIONS

The above discussion has focused on governmental units at the state, tribal, county, city, or local levels. However, it should not be assumed that these jurisdictions bear the full responsibility for extending the Framework. There are at least two roles for private industry and nonprofit organizations in the creation of an extended Framework:

1. Performing the actual data capture and database creation under contract to governmental units; and
2. Involvement in consortia of private firms, nonprofit organizations, and governmental units in collecting and maintaining necessary data.

Examples can be cited at all levels of government of the use of private industry to convert analog geographic information into digital form. This arrangement is likely to continue: it does not make economic sense for governmental units in most instances to carry out the conversion of existing data, since this is a large one-time operation that can be carried out efficiently in the private sector. On the other hand, if the governmental unit does some comprehensive planning that includes provisions for maintenance prior to conversion, it makes economic sense for governmental operations to perform the maintenance and update functions. Unfortunately, to date much conversion has been accomplished without sufficient concern for maintenance and update, and it can therefore be expected that private firms and nonprofit organizations will also be needed for the initial update of the converted analog data.

More important to the long-term maintenance and health of Framework data is the recognition by private industry that its future lies in providing services for individuals and firms that utilize the extended NSDI. Once that realization occurs, we will find that it is in the best economic interests of industry, nonprofit organizations, and government to form consortia to ensure continued availability of the data needed for a robust extended NSDI. The long-term role of private industry in an extended NSDI is to provide spatial data services to consumers, including individuals, corporations, governmental units at all levels, and nonprofit organizations. The Committee envisions the extended NSDI data to be a public asset. Ideally, the creation and development of useful information from these data, provided by service-oriented businesses, will constitute a lucrative marketplace. The private sector will also continue to have a major role in developing and maintaining the data. It will also provide valuable software tools that will enable communities to better serve their citizens.

We are fortunate in the United States that some of the leaders in the geospatial business community are already adopting this mode of thinking and implementation. The New York State Office of Technology has a Data Sharing Cooperative Agreement that recognizes the benefits of data remaining in the public domain (distributed at no more than the cost of reproduction and shipping), enabling access to those data for all users, including value-added information-service marketing firms. There are certainly firms that still try to generate profits by selling digital data that are available to anyone. Once they understand the future, these firms can easily migrate to providing a useful service by enhancing a customer's use of digital data rather than by selling the data themselves.

The creation and maintenance of spatial data represents a substantial investment by a community. It must be recognized that there is a great disparity among local governments across the country in their ability to support an extended framework from both technical and financial perspectives. While many communities have devised creative ways to finance such systems, others will never be in a position to do so. Regional or even statewide consortia will be required to develop a consistent level of spatial data. Furthermore, in some parts of the country, mechanisms such as Geographic Information Block Grants will be required to overcome this "spatial digital divide."

5

Conclusions and Recommendations

Over the past seven years, the programs of the Federal Geographic Data Committee have been very successful in several respects. They have promoted the concepts and objectives of the National Spatial Data Infrastructure, and helped to ensure that the NSDI is a familiar acronym among government agencies at all levels, in academic environment, and among the private sector. They have initiated the National Geospatial Data Clearinghouse, and recruited a substantial number of servers to its transparent network. They have also promulgated standards, including 16 that have been endorsed by the community. These include the Content Standards for Digital Geospatial Metadata, a major contribution to the FGDC's effort to promote greater sharing of geospatial data and less redundancy in its production.

The various partnership programs analyzed in this report have contributed significantly to this effort. All states except North Dakota have received funding from at least one program, and a large number of partnerships have been initiated during the process of competing for these awards, and sustained by the federal funding. We conclude that the programs have succeeded in their role of launching the NSDI, and spreading awareness of it throughout the geospatial data community.

The various programs have also played a significant role in seeding NSDI activities in smaller states, smaller agencies, and organizations with minimal resources. In this respect they have helped to "level the playing field," and to ensure that the benefits of the NSDI are available to all. However, it is the view of the committee that small-scale efforts designed to attract attention to the NSDI need to give way to larger-scale production efforts. Some research indicates that fewer than half of the local government entities in the United States are even aware of the meaning of NSDI. This suggests that there is a great deal of work remains to be done. The FGDC should be encouraged to get the word out through as many venues as possible and provide clear examples of how to participate and the benefits that can be gained.

This study evaluated the partnership programs against four goals. One of these, improving access to geospatial data, has been greatly aided by the development of the Internet and World Wide Web, and the FGDC was quick to exploit the advantages of these technologies in the development of the National Geospatial Data Clearinghouse. We conclude that the programs have been very successful in achieving this third goal.

However, with respect to the other goals of the specific FGDC partnership programs, we find little evidence that these programs have reduced redundancy in geospatial data creation and maintenance, reduced the costs of geospatial data creation and maintenance, or improved the accuracy of the geospatial data used by the broader community. For all three goals, little evidence has been found to demonstrate conclusively that the concept of the NSDI and its furtherance through partnerships has had any dramatic impact on overcoming the significant institutional barriers that inhibit the development and maintenance of spatial data. Without such evidence, we fear that the momentum established as a result of the missionary efforts during these seven years will dissipate, and that the NSDI will fail to achieve its promise.

In our investigations, we looked for ways of assessing the impacts of the partnership programs using objective indicators and metrics. We found indicators of the level of interest in the NSDI at the state level, as discussed in Chapter 3. But we found a lack of procedures in the FGDC for long-term monitoring of the progress of NSDI. Such procedures would be of great value in assessing whether the NSDI program succeeds in moving beyond the missionary phase, and in arguing for future funding allocations. Accordingly, the committee recommends that the FGDC develop metrics that can be used to monitor long-term progress in the adoption of the principles and programs of the NSDI among agencies at all levels of govern ment, academia, and the private sector. The Committee advocates adoption of a funding formula that provides resources to all participants on a non-competitive basis, coupled with grants of sufficient size and duration to achieve expected outcomes. In addition, the committee recommends that funding should be directed at projects that are of a sufficient scale to provide well-designed empirical tests of the hypotheses underlying the NSDI goals, and should allow for adequate documentation and dissemination of results.

In our discussions, we were struck by the many forms of partnership that have emerged over the past seven years. Partnerships exist at all levels of government, and involve all types of organizations and agencies. Only a small proportion of them have received substantial funding from the FGDC programs, and in those cases the amount of funding provided was comparatively small relative to the total resources available to the partnership. It is difficult to see the complete picture if one focuses too much on the FGDC's programs, and difficult to set these in the correct context. The Committee recommends that future partnership programs initiated by the FGDC should be conceived in the context of all relevant partnership programs, and should be designed to augment and leverage them to achieve maximum impact.

The NSDI is at a critical juncture in its evolution. The FGDC continues to play the lead role of federal coordination. The efforts of the working groups and subcommittees have resulted in important dialog among the stakeholders and standards for the definition of different data components are emerging. At the same time, a new organization such as the GeoData Alliance could radically change the institutional setting for the promotion of the NSDI. The new initiative by the OMB demonstrates the importance of spatial data and recognizes that the Federal government has a limited role in its actual maintenance. We find it encouraging and surprising that the OMB initiative has been rapidly adopted as a useful umbrella for coordinating data sharing efforts at a variety of regional levels. The activities of these I-Teams must be carefully analyzed to determine whether a "bottom-up" model can be successful. We are also at an interesting stage in technological development that is driving a robust private sector. Commercial remote sensing satellites are providing data that are suitable for extraction of some urban features (e.g.,

LIDAR (LIght Detection And Ranging), IKONOS, and SPOT data, coupled with GPS, are providing enormous improvements in the capture of digital terrain data. The wealthier local governments are making substantial investments in spatial data to support more responsive and accountable form of services to the taxpayer. Commercial demand for street centerlines and postal code data is accelerating at the same time the Bureau of Census is releasing the 2000 census data and is contemplating the need for modernization of its TIGER database. All of these factors reinforce the Committee's original view of a national need for a robust NSDI that is in the public domain. The Committee also appreciates that a successful NSDI must address the need for business plans that encourage private sector involvement and local government investments.

References

Corbis, 2001. Online; available at http://store.corbis.com/; accessed February, 2001.

Cowen, D.J., and J.Jensen, 1998. Extraction and modeling of urban attributes using remote sensing technology. Pp. 164–188 in People and Pixels: Linking Remote Sensing and Social Science. Washington, DC: National Academy Press.

CSDGM [Content Standards for Digital Geospatial Metadata], 2001. Online; available at http://www.fgdc.gov/metadata/contstan. html; accessed February 2001.

ESRI [Environmental Systems Research Institute], 2001a. The Geography Network and the NSDI, Whitepaper. Online; available at http://www.esri.com/library/whitepapers/pdfs/GN_NDSI.pdf; accessed January 2001.

ESRI [Environmental Systems Research Institute], 2001b. Tennessee Begins First-of-its-Kind GIS Basemapping Project. ArcNews 23(1):1,3.

Federal Register, 1994. Executive Order 12906: Coordinating Geographic Data Acquisition and Access. The National Spatial Data Infra-structure 59:17671–17674.

FGDC [Federal Geographic Data Committee], 1994. National Spatial Data Infrastructure Competitive Cooperative Agreements Program. Online; available at http://www.fgdc.gov/publications/documents/cooperativeagreements/ fundingprograms/94cap.pdf; accessed April 2001.

FGDC [Federal Geographic Data Committee], 1995. Development of a National Digital Geospatial Data Framework. Framework Working Group. Online; available at http://www.fgdc.gov/framework/framdev.html; accessed April 1995.

REFERENCES

FGDC [Federal Geographic Data Committee], 1996. National Spatial Data Infrastructure. 1996 Framework Demonstration Project Program (FDPP). Federal Geographic Data Committee. Online; available at http://www.fgdc.gov/ publications/documents/cooperativeagreements/ funding programs/96fdpp.pdf; accessed April 2001.

FGDC [Federal Geographic Data Committee], 1997'a. Framework: Introduction and Guide. Federal Geographic Data Committee. Online; available at http://www.fgdc.gov/frame work/ framework introguide/; accessed March 2001.

FGDC [Federal Geographic Data Committee], 1997b. Impacts of the NSDI Competitive Cooperative Agreements Programs. Unpublished report, Washington, DC.

FGDC [Federal Geographic Data Committee], 1998. 1998 National Spatial Data Infrastructure Funding Programs. Federal Geographic Data Committee, Washington, D.C.

FGDC [Federal Geographic Data Committee], 2001. National Spatial Data Infrastructure Community Demonstration Projects—Final Report. Online; available at http://www.fgdc.gov/nsdi/docs/cdp/; accessed February 2001.

GeoData Alliance, 2000. GeoData Alliance Final Drafting Team Report, September 18, 2000. Online; available at http:// www.geoall.net/downloads/GDA_Final_DT_Report.pdf; accessed February 2001.

Harvey, F., 2001. NSDI from the trenches: local governments surveyed. Geospatial Solutions 11 (5):38–40.

Longley, P.A., M.F.Goodchild, D.J Maguire, and D.W.Rhind, 2001. Geographic Information Systems and Science. Chichester: John Wiley and Sons.

Mayo, J., 1985. The Evolution of Information Technologies, pp. 7–33 In B.Guile (ed.), Information Technologies and Social Transformation. Washington, DC: National Academy Press.

NAPA [National Academy of Public Administration], 1998. Geographic Information for the 21st Century: Building a Strategy for the Nation Washington, DC: NAPA.

REFERENCES

NGDC [National Geospatial Data Clearinghouse], 2001. Online; available at http://www.fgdc.gov/clearinghouse/clearing house.html; accessed February 2001.

NIST [National Institute of Science and Technology], 1994. Federal Information Processing Standard 173-1: Spatial Data Transfer Standard. Department of Commerce. Online; available at http://mcmcweb.er.usgs.gov/sdts/standard.html; accessed March 2001.

NPR [National Partnership for Reinventing Government], 1993. Online; available at http://www.npr.gov/whoweare/historyof npr.html; accessed January 2001.

NRC [National Research Council], 1980. Need for a Multipurpose Cadastre. Washington, DC: National Academy Press, 112 pp.

NRC [National Research Council], 1993. Toward a Coordinated Spatial Data Infrastructure for the Nation. Washington, DC: National Academy Press, 171 pp.

NRC [National Research Council], 1994. Promoting the National Spatial Data Infrastructure Through Partnerships. Washington, DC: National Academy Press, 113 pp.

NRC [National Research Council], 1995. A Data Foundation for the National Spatial Data Infrastructure. Washington, DC: National Academy Press, 45 pp.

OGC [Open GIS Consortium], 2001. Online; available at http://www.opengis.org; accessed February 2001.

OMB [Office of Management and Budget], 1990. Circular A-16. Coordination of Surveying, Mapping, and Related Spatial Data Activities. Washington, DC.

OMB [Office of Management and Budget], 2000. OMB Information Initiative Collection Information in the Information Age. Washington, DC.

Qpass, 2001. Connected Commerce. It's Coming Together. Online; available at http://www.qpass.com; accessed March 2001.

Rogers, E., 1995. The Diffusion of Innovations, 4th Edition. New York: Free Press.

Somers, R., 1999. Framework Data Survey: Preliminary Report. Geo Info Systems, September Supplement.

REFERENCES

Tulloch, D.L., 1999. Theoretical model of multipurpose land information systems development. Transactions in GIS 3(3):259– 283.

UCGIS [University Consortium for Geographic Information Science], 2000a. Request for Proposals: Evaluation of FGDC's NSDI Grants Program UCGIS RFP 0001. Online; available at http:/ /www.spatial.maine.edu/~max/UCGIS/0001RFP.html; accessed March 2001.

UCGIS [University Consortium for Geographic Information Science], 2000b. Cross-Cutting the UCGIS Challenges in Geographic Information Science: Reviews of Essential Progress and Vision in Major Application Domains . URISA Journal 12(2, Special Issue): 4–93.

UCGIS [University Consortium for Geographic Information Science], 2001. Online; available at http://www.ucgis.org; accessed February 2001.

UDRP [Urban Dynamics Research Program], 2001. Online; available at http://edcwww2.cr.usgs.gov/umap/umap.html; accessed February 2001.

Utah Geographic Information Systems Advisory Council, 2001. Utah Framework Implementation Plan. Online; available at http://agrc.its.state.ut.us/i_team/paradigm.pdf; accessed September 2001.

Acronyms

CAP	Cooperative Agreements Program
CFIP	Community-Federal Information Partnerships
CRADAs	Cooperative Research and Development Agreements
CSDGM	Content Standards for Digital Geospatial Metadata
EPA	Environmental Protection Agency
ESRI	Environmental Systems Research Institute
FDPP	Framework Demonstration Project Program
FGDC	Federal Geographic Data Committee
FIPS	Federal Information Processing Standard
GI	Geographic Information
GIS	Geographic Information System
GPS	Global Positioning System
GSDI	Global Spatial Data Infrastructure
LIDAR	Light Detection And Ranging
LSDI	Local Spatial Data Infrastructure
MSC	Mapping Science Committee
NACo	National Association of Counties
NAD 83	North American Datum, 1983
NAPA	National Academy of Public Administration
NASA	National Aeronautics and Space Administration
NAVD 88	North American Vertical Datum, 1988
NERR	National Estuarine Research Reserve
NGDC	National Geospatial Data Clearinghouse
NGO	Non-government Organization
NIST	National Institute of Science and Technology
NOAA	National Oceanic and Atmospheric Administration
NPR	National Partnership for Reinventing Government
NRC	National Research Council

ACRONYMS

NRCS	Natural Resource Conservation Service
NSDI	National Spatial Data Infrastructure
NSGIC	National States Geographic Information Council
OGC	Open GIS Consortium
OMB	Office of Management and Budget
PLSS	Public Lands Survey System
SCSD	Street Centerline Spatial Database
SDTS	Spatial Data Transfer Standard
SPC	State Plane Coordinate
SSDI	State Spatial Data Infrastructure
TIGER	Topologically Integrated Geographic Encoding and Referencing system
UCGIS	University Consortium for Geographic Information Science
UDRP	Urban Dynamics Research Program
USGS	U.S. Geological Survey